Novel Photoactive Materials

Special Issue Editor

Maria Vittoria Diamanti

MDPI • Basel • Beijing • Wuhan • Barcelona • Belgrade

Special Issue Editor
Maria Vittoria Diamanti
Politecnico di Milano
Italy

Editorial Office
MDPI
St. Alban-Anlage 66
4052 Basel, Switzerland

This is a reprint of articles from the Special Issue published online in the open access journal *Materials* (ISSN 1996-1944) in 2018 (available at: https://www.mdpi.com/journal/materials/special_issues/novel_photoactive_materials)

For citation purposes, cite each article independently as indicated on the article page online and as indicated below:

LastName, A.A.; LastName, B.B.; LastName, C.C. Article Title. *Journal Name* **Year**, *Article Number*, Page Range.

ISBN 978-3-03897-650-9 (Pbk)
ISBN 978-3-03897-651-6 (PDF)

Cover image courtesy of unsplash.com user Dmitry Bayer.

© 2019 by the authors. Articles in this book are Open Access and distributed under the Creative Commons Attribution (CC BY) license, which allows users to download, copy and build upon published articles, as long as the author and publisher are properly credited, which ensures maximum dissemination and a wider impact of our publications.
The book as a whole is distributed by MDPI under the terms and conditions of the Creative Commons license CC BY-NC-ND.

Novel Photoactive Materials

Contents

About the Special Issue Editor

Maria Vittoria Diamanti, Associate Professor: Maria Vittoria Diamanti completed a Ph.D. in Materials Engineering at Politecnico di Milano in 2009. She is currently an Associate Professor at the Department of Chemistry, Materials and Chemical Engineering "Giulio Natta" at Politecnico di Milano, where she teaches cementitious and ceramic materials principles and durability to Bachelor and Master's students of materials engineering and nanotechnologies. In 2007, Maria Vittoria was awarded a Roberto Rocca fellowship which allowed her to spend a period as a visiting student at the Massachusetts Institute of Technology (MIT). In 2012 she was appointed a visiting professor at the Masdar Institute of Technology, Abu Dhabi with the LENS group (Laboratory for Energy and Nano Sciences). Her research interests mostly relate to the production, characterization and applications of nanostructured titanium oxides, either grown by electrochemical techniques on titanium and its alloys, or as nanoparticles or sol-gel applied to different substrates as coating or in bulk. The versatility of titanium dioxide has allowed her to get expertise in diverse fields—from photocatalysis for wastewater treatment and air quality improvement, to bioengineering, jewelry, corrosion resistant surfaces, the use of memristive materials for electronics, self-cleaning and the use of photoactive materials as a complement to building technologies—and to use a multidisciplinary approach to the production, characterization and application of these oxides. Maria Vittoria is also involved in research activities on the corrosion resistance of steel alloys in particular environments, such as chloride containing environments, concrete and new electrolytes, such as ionic liquids.

Editorial

Special Issue: Novel Photoactive Materials

Maria Vittoria Diamanti

Department of Chemistry, Materials and Chemical Engineering "G. Natta" Politecnico di Milano,
Via Mancinelli 7, 20131 Milan, Italy; mariavittoria.diamanti@polimi.it; Tel.: +39-02-2399-3137

Received: 28 November 2018; Accepted: 12 December 2018; Published: 15 December 2018

Abstract: Photoactivity represents the ability of a material to activate when interacting with light. It can be declined in many ways, and several functionalities arising from this behavior of materials can be exploited, all leading to positive repercussions on our environment. There are several classes of effects of photoactivity, all of which have been deeply investigated in the last few decades, allowing researchers to develop more and more efficient materials and devices. The special issue "Novel Photoactive Materials" has been proposed as a means to present recent developments in the field; for this reason the articles included touch different aspects of photoactivity, from photocatalysis to photovoltaics to light emitting materials, as highlighted in this editorial.

Keywords: photocatalysis; photovoltaics; organic light emitting diodes (OLEDs); TiO_2; ZnO; density functional theory (DFT)

Generally speaking, photoactive materials belong to the huge field of photonics, where materials that actively interact with light are tuned and optimized to achieve effects such as light emission (LEDs and lasers, just to name the most common ones) or light detection, with related signal amplification (e.g., in photomultipliers) and processing operations. Alternatively, they can be used to develop light-sensitive circuits and switches (such as with photoresistors), or more generally, to convert light into an electrical signal (i.e., to build photodiodes). All of these materials and devices are not only subject of research, but already find diffuse applications in our everyday life.

In particular, the latter case of photodiodes includes devices with huge technological importance in the field of clean energy production, i.e., photovoltaic devices, whose functioning principle is based on the generation of electrical current as a consequence of sunlight absorption, and specifically on the jump of electrons from the valence band to the conduction band when the photoactive material is irradiated with light having energy equal or higher the material bandgap. A similar mechanism, where the photo-generated electrons and holes are exploited in a different way to accelerate chemical reactions, is related to environmental cleanup, as a result of photocatalytic degradation of pollutants in gas and liquid phase. This is currently exploited in air purification devices and water remediation reactors. This same effect can be declined in a similar manner to achieve the so-called self-cleaning effect, where the decomposition of contaminants couples with a further peculiarity of photoinduced superhydrophilicity, which finds relevant applications in building materials due to lower requirements for maintenance and potentially higher durability of self-cleaning materials.

These effects share a common point, that is, the interaction of a material with light, although many different materials are taken into account depending on the effect desired—from elemental semiconductors like silicon, to more complex compounds like CdTe or GaAs, to metal oxides like TiO_2 and ZnO. Given the broadness of the field, a huge number of works fall within this topic, and new areas of discovery are constantly explored.

The special issue "Novel Photoactive Materials" has been proposed as a means to present recent developments in the field, and for this reason the eleven articles included touch different aspects of photoactivity: a brief summary of the articles' content is given in the following.

The article by Kim et al. [1] reported on the synthesis of 8-hydroxyquinolinato aluminum microwires showing nanoscale photoluminescence properties, with applications as fluorophore in organic light emitting diodes (OLEDs): the enhanced crystal structure and architecture of hexagonal microwires allowed the improvement of optical properties in the material solid state.

In two articles, the subject of study was photoactive materials for solar cells. In particular, in their feature article, Salado et al. [2] presented a study on the role of interfaces on the performances of multilayer organo-halide perovskite solar cells, with specific reference to the electron selective layer. Indeed, its morphology is a key parameter in determining its performance in extracting charges from the perovskite layer, since energy losses may play a relevant role at this interface. In this respect, a mesoporous or nanostructured interfaces ensured better performance and stability with respect to planar ones, and the recombination mechanism was switched from a mixed mechanism to a trap-limited one. On the other hand, Liu et al. [3] proposed a study on methylammonium lead iodide as light absorber in hybrid solar cells, investigating its geometry, electronic structure, thermodynamic, and mechanical property by density functional theory (DFT) first-principles calculations in order to understand its stability, which is currently a major concern for its application. Caesium (Cs) doping was also taken into account, and the authors demonstrated that its presence can enhance the compound thermodynamic and mechanical stability with no or little influence on its conversion efficiency.

Photocatalysis was the main theme of the other eight articles. A first feature paper by Brzezińska et al. [4] investigated the photocatalytic performance of Cu-ZnO heterojunction co-catalysts, using several zinc oxide formulations obtained with different preparation methods. Photocatalysis tests, focused on the decomposition of 4-chlorophenol in water solution under simulated solar light, allowed the researchers to select the best ZnO synthesis method. Moreover, the possibility to control the oxidation state of the Cu co-catalyst with simple room-temperature treatments was demonstrated. Another feature article by Sanabria Arenas et al. [5] presented a comparison between gas and liquid phase photocatalytic activity of several titanium dioxide (TiO_2) films grown by anodic oxidation, presenting different morphology. Toluene and rhodamine B (RhB) degradation were selected as model gas and liquid phase reactions, respectively. A strong influence of oxide surface area and crystallinity was observed in both reactions, especially in the gas phase, where the reaction was largely more sensitive to oxide characteristics than the one in liquid phase, showing a less resilient nature of the RhB dye with respect to gas phase toluene. Titanium dioxide was the subject of studies by Cheng et al. [6], Janek et al. [7] and Roveri et al. [8]. The former article proposed the sol-gel production of nitrogen and iron co-doped TiO_2 catalysts, which were then tested in the degradation of an organic dye, acid orange 7 (AO7). The doping allowed the researchers to obtain high visible light activity, which was further accentuated by choosing a suitably low pH of the AO7 solution, namely pH 3 [6]. In the study by Janek [7], DFT calculations were coupled to physico-chemical analyses to study titanium(IV) oxo-clusters of different formula, allowing the researchers to highlight the influence of substitutes on the measured material band gap and photocatalytic activity, which was estimated by using the degradation of methylene blue (MB) dye as a model reaction. Roveri et al. [8] proposed the use of nanocomposite coatings based on alkylalkoxysilane matrices and TiO_2 nanoparticles as protective layers on porous stone substrates, and their performances were evaluated to assess the influence of titanium dioxide on the coatings durability, as well as water absorption properties and photocatalytic behavior, studied in the degradation of RhB. The composite coatings allowed the researchers to reach high photocatalytic efficiency while maintaining good protective performances, with only slight effects on the coating durability.

Several other studies involved the evaluation of different material's photocatalytic activity in the degradation of organic dyes. Tseng et al. [9] chose the degradation of MB to evaluate the photocatalytic activity of their Ag_3PO_4 microparticles, which were also tested in the degradation of phenol, both under low-power white-light light-emitting-diode. The high activity obtained was correlated with the microparticles' large surface area, low recombination rate, and the high charge separation efficiency; further characterization of their photocatalytic activity was carried out in the treatment of environmental water samples polluted with MB, methyl red, acid blue 1, and RhB. Ding et al. [10] used Congo red as model dye to evaluate the photocatalytic efficiency of $Bi_2O_2(CO_3)_{1-x}S_x$ synthesized by chemical bath precipitation. The control of S content in the precipitation process was observed to play a major role in the material properties, as it broadened the optical absorption of the bismuth oxide-based catalyst towards the visible light region.

Finally, Palma et al. [11] investigated the possibility to exploit compost-derived bio-based substances to prepare auxiliary hybrid magnetic nanoparticles (HMNPs) for water treatment processes, as a means for an easier recovery of the nanoparticles dispersed in the effluent; their iron-based HMNPs were then used in the photo-Fenton degradation of caffeine in combination with hydrogen peroxide and UV light, whose role was that of generating hydroxyl radicals. The added value of the HMNPs was an increase in the pH of the caffeine solution required to obtain its degradation, and the organic coverage coming from compost also improved the overall degradation reaction.

Conflicts of Interest: The author declares no conflict of interest.

References

1. Kim, S.; Kim, D.; Choi, J.; Lee, H.; Kim, S.; Park, J.; Park, D. Growth and Brilliant Photo-Emission of Crystalline Hexagonal Column of Alq3 Microwires. *Materials* **2018**, *11*, 472. [CrossRef] [PubMed]
2. Salado, M.; Calió, L.; Contreras-Bernal, L.; Idígoras, J.; Anta, J.; Ahmad, S.; Kazim, S. Understanding the Influence of Interface Morphology on the Performance of Perovskite Solar Cells. *Materials* **2018**, *11*, 1073. [CrossRef] [PubMed]
3. Liu, D.; Li, S.; Bian, F.; Meng, X. First-Principles Investigation on the Electronic and Mechanical Properties of Cs-Doped $CH_3NH_3PbI_3$. *Materials* **2018**, *11*, 1141. [CrossRef] [PubMed]
4. Brzezińska, M.; García-Muñoz, P.; Ruppert, A.; Keller, N. Photoactive ZnO Materials for Solar Light-Induced CuxO-ZnO Catalyst Preparation. *Materials* **2018**, *11*, 2260. [CrossRef] [PubMed]
5. Sanabria Arenas, B.; Strini, A.; Schiavi, L.; Li Bassi, A.; Russo, V.; Del Curto, B.; Diamanti, M.; Pedeferri, M. Photocatalytic Activity of Nanotubular TiO_2 Films Obtained by Anodic Oxidation: A Comparison in Gas and Liquid Phase. *Materials* **2018**, *11*, 488. [CrossRef] [PubMed]
6. Cheng, H.; Chen, S.; Yang, S.; Liu, H.; Lin, K. Sol-Gel Hydrothermal Synthesis and Visible Light Photocatalytic Degradation Performance of Fe/N Codoped TiO_2 Catalysts. *Materials* **2018**, *11*, 939. [CrossRef]
7. Janek, M.; Radtke, A.; Muzioł, T.; Jerzykiewicz, M.; Piszczek, P. Tetranuclear Oxo-Titanium Clusters with Different Carboxylate Aromatic Ligands: Optical Properties, DFT Calculations, and Photoactivity. *Materials* **2018**, *11*, 1661. [CrossRef] [PubMed]
8. Roveri, M.; Gherardi, F.; Brambilla, L.; Castiglioni, C.; Toniolo, L. Stone/Coating Interaction and Durability of Si-Based Photocatalytic Nanocomposites Applied to Porous Lithotypes. *Materials* **2018**, *11*, 2289. [CrossRef]
9. Tseng, C.; Wu, T.; Lin, Y. Facile Synthesis and Characterization of Ag_3PO_4 Microparticles for Degradation of Organic Dyestuffs under White-Light Light-Emitting-Diode Irradiation. *Materials* **2018**, *11*, 708. [CrossRef] [PubMed]
10. Ding, J.; Wang, H.; Xu, H.; Qiao, L.; Luo, Y.; Lin, Y.; Nan, C. Synthesis and Broadband Spectra Photocatalytic Properties of $Bi_2O_2(CO_3)_{1-x}S_x$. *Materials* **2018**, *11*, 791. [CrossRef] [PubMed]
11. Palma, D.; Bianco Prevot, A.; Brigante, M.; Fabbri, D.; Magnacca, G.; Richard, C.; Mailhot, G.; Nisticò, R. New Insights on the Photodegradation of Caffeine in the Presence of Bio-Based Substances-Magnetic Iron Oxide Hybrid Nanomaterials. *Materials* **2018**, *11*, 1084. [CrossRef] [PubMed]

© 2018 by the author. Licensee MDPI, Basel, Switzerland. This article is an open access article distributed under the terms and conditions of the Creative Commons Attribution (CC BY) license (http://creativecommons.org/licenses/by/4.0/).

Article

Growth and Brilliant Photo-Emission of Crystalline Hexagonal Column of Alq₃ Microwires

Seokho Kim [1,†], Do Hyoung Kim [1,†], Jinho Choi [1], Hojin Lee [1], Sun-Young Kim [1], Jung Woon Park [1] and Dong Hyuk Park [1,2,*]

[1] Department of Applied Organic Materials Engineering, Inha University, Incheon 22212, Korea;
    seokho@inha.edu (S.K.); gg1236@inha.edu (D.H.K.); jinho@inha.edu (J.C.); hojin@inha.edu (H.L.);
    22161060@inha.edu (S.-Y.K.); jungwoon@inha.edu (J.W.P.)
[2] Department of Chemical Engineering, Inha University, Incheon 22212, Korea
*   Correspondence: donghyuk@inha.ac.kr; Tel.: +82-32-860-7496
†   These authors contributed equally.

Received: 28 February 2018; Accepted: 19 March 2018; Published: 22 March 2018

Abstract: We report the growth and nanoscale luminescence characteristics of 8-hydroxyquinolinato aluminum (Alq₃) with a crystalline hexagonal column morphology. Pristine Alq₃ nanoparticles (NPs) were prepared using a conventional reprecipitation method. Crystal hexagonal columns of Alq₃ were grown by using a surfactant-assisted self-assembly technique as an adjunct to the aforementioned reprecipitation method. The formation and structural properties of the crystalline and non-crystalline Alq₃ NPs were analyzed with scanning electron microscopy and X-ray diffraction. The nanoscale photoluminescence (PL) characteristics and the luminescence color of the Alq₃ single NPs and their crystal microwires (MWs) were evaluated from color charge-coupled device images acquired using a high-resolution laser confocal microscope. In comparison with the Alq₃ NPs, the crystalline MWs exhibited a very bright and sharp emission. This enhanced and sharp emission from the crystalline Alq₃ single MWs originated from effective π-π stacking of the Alq₃ molecules due to strong interactions in the crystalline structure.

Keywords: organometal; Alq₃; photoluminescence; crystallinity; surfactant; confocal microscope

---

1. Introduction

Organic semiconductors have emerged as a major class of plausible candidates for achieving low-cost, flexible and high-efficiency electronic devices [1,2]. Intensive research is currently carried out worldwide using crystalline organic semiconductors as active materials for optoelectronic devices because of the semiconducting characteristics and the excellent photo-induced charge creation of these molecules [3]. Single-crystalline organic small molecules are among the best recognized materials, as there are no grain boundaries and few charge trapping sites [4,5]. Among the various organic small molecules, organometallics have attracted much attention in recent years owing to their unique optical and electrical properties and their various applications [6,7]. Organometallic molecules are composed of a central metal surrounded by an organic conjugated structure. These combinations of metals and organic compounds generally furnish optoelectronic properties. These properties can be controlled by altering structural features such as the crystallinity and molecular aggregation, or by chemical processes (e.g., doping or substitution). One interesting phenomenon of organometallic molecules is their light-induced or electrically-driven luminescence. Many efforts have been made to improve luminous efficiency, including the use of fluorophores.

The best known of these fluorophores—8-hydroxyquinolinato aluminum (Alq₃)—exhibits excellent optical properties that can be exploited in optoelectronic devices, such as its use as the active light emissive layer of organic light emitting diodes (OLEDs) [8,9]. Generally, Alq₃ dissolved in an organic solvent emits bright yellow-green under ultraviolet (UV) irradiation. However, in the solid-state, amorphous Alq$_3$ exhibits very low brightness under the same UV irradiation conditions due to random stacking in the molecules. Growth of this architecture for application to a variety of optoelectronic devices can thus be problematic. Recent studies have reported various methods for the growth of regular and crystalline Alq$_3$ morphologies, such as hexagonal columns, using vapor deposition or solvent treatment. These approaches lead to enhanced optical properties due to enhanced crystallinity or molecular orientation [10,11].

In recent years, organic small molecule crystals have attracted increasing research interest due to their potential use in optoelectronics and photonics. Much research effort has been devoted to molecule design and crystal growth to achieve the desired function [12,13]. The electronic and optical properties of organometallics are fundamentally different from those of inorganic metals and semiconductors due to the weak intermolecular van der Waals forces. Therefore, understanding and controlling the arrangement of molecules in the solid-state are fundamental issues for obtaining the desired chemical and physical properties. These non-covalent intermolecular interactions, such as hydrogen bonding and π-π stacking, can strongly influence the final packing structure [14,15]. Moreover, there are relatively few reports on the optical characteristics of single units in the solid-state. Thus, this study focuses on the nanoscale photoluminescence (PL) characteristics of Alq$_3$ crystals in the solid-state in an attempt to examine the changes in the structural and optical properties upon crystallization [16,17].

Herein, we report the growth and nanoscale luminescence characteristics of single-crystalline Alq$_3$ microwires (MWs) as shown in Figure 1. Crystalline Alq$_3$ MWs with a hexagonal column arrangement can be grown via surfactant-assisted self-assembly using sodium dodecyl sulfate (SDS) [18]. The surfactant induces the formation of micelles in deionized (D.I.) water and the Alq$_3$ molecules penetrate the micelle and initiate the nucleation reaction. Because organometallic molecules are not amenable to water due to their inherent polarity, they penetrate the micelle to reduce surface energy. After nucleation, the Alq$_3$ molecules act as seeds for growth of the hexagonal 1D MWs. We demonstrate that hexagonal column crystal Alq$_3$ has a much brighter emission profile than randomly oriented Alq$_3$ nanoparticles in a single unit.

Figure 1. Schematic illustration of surfactant-assisted organic crystal growth using reprecipitation method.

2. Materials and Methods

Sample preparation: Alq$_3$ (C$_{27}$H$_{18}$AlN$_3$O$_3$, purity 99.995%) and SDS (CH$_3$(CH$_2$)$_{11}$OSO$_3$Na, purity 99.0%) were purchased from Sigma Aldrich (St. Louis, MI, USA). Alq$_3$ was dissolved in tetrahydrofuran (THF) to give a concentration of 4 mg·mL^{-1}. The Alq$_3$ solution was heated to 50 °C and homogenized using a magnetic stirrer for complete dissolution; SDS (4 mg·mL^{-1}) was prepared in D.I. water.

The SDS solution was poured into a 20-mL vial and vigorously magnetically stirred on the hot-plate. The Alq_3 solution was added and forcefully ejected into the SDS solution using a micro-pipette. The vial was capped and stirring was continued at a high rpm for 15 min. The mixture was kept in an oven at 50 °C for 40 h. Another solution was prepared with the same formulation used above, without the addition of SDS.

Measurement: The surface morphology of the Alq_3 MWs was analyzed using a field-emission scanning electron microscope (SEM; Hitachi, Tokyo, Japan, SU-8010) at an acceleration voltage of 15 kV. The powder X-ray diffraction (XRD; X'Pert Powder Diffractometer, PANalytical) patterns were captured at a voltage of 40 kV and a current of 40 mA using Cu-Kα radiation (λ = 1.540 Å). The scan rate was $0.02° \cdot s^{-1}$ and the 2θ range was from 2° to 60°. The luminescence color charge-coupled device (CCD) images of Alq_3 were acquired with an AVT Marlin F-033C (λ_{ex} = 435 nm) instrument. To compare the brightness (i.e., luminescence intensity) of the CCD images of the particles and crystals of Alq_3, the irradiation time was fixed at 0.1 s. Laser confocal microscope (LCM) PL spectra were acquired with a homemade LCM instrument [19–22]. The 405 nm line of an unpolarized diode laser was used for the LCM PL excitation. The organic crystals were placed on a glass substrate, which was mounted on the xy-stage of the confocal microscope. An oil-immersion objective lens (N.A. of 1.4) was used to focus the unpolarized laser light on the crystal. The spot size of the focused laser beam on the sample was approximately 200 nm. The scattered light was collected with the same objective lens and the excitation laser light was filtered out with a long-pass edge-filter (Semrock, Rochester, NY, USA). The red-shifted PL signal was focused onto a multimode fiber (core size = 100 μm) that acted as a pinhole for the confocal detection. The other end of the multimode fiber was connected to the photomultiplier tube for acquisition of the PL image, or the input slit of a 0.3 m long monochromator equipped with a cooled charge coupled device for acquisition of the PL spectra. Solid-state PL measurements could therefore be performed at the nanometer scale. The laser power incident on the sample and the acquisition time for each LCM PL spectrum were fixed at 500 μW and 1 s, respectively, for all confocal PL measurements.

3. Results

3.1. Morphological Analysis

The formation of the Alq_3 NPs and the crystal hexagonal column of Alq_3 MWs was visually observed through SEM imaging, as shown in Figure 2a,b,. The mean diameter of the Alq_3 NPs was estimated to be 150 (±50) nm, as shown in Figure 2a. The NPs grew randomly into sphere-like plates or particles because the Alq_3 molecules aggregated randomly in the solvent. Thus, there was no specific shape or direction. When a surfactant was employed along with the reprecipitation process, crystalline hexagonal columns were formed with growth in only one specific direction. This is because the SDS (as a surfactant) micelles assist the nucleation of Alq_3 during reprecipitation. The Alq_3 nuclei can act as seeds for growth of the hexagonal column. We found that the cross-section of the Alq_3 single crystal was hexagonal, based on the tilted SEM image, with a length of 5–30 μm depending on the growth conditions, such as the time, solvent temperature and concentration—as shown in Figure 2b. These results suggest that the SDS surfactant transformed the amorphous non-crystalline Alq_3 NPs into a crystalline phase with growth in a 1D manner, resulting in the hexagonal column MWs. The side-view and top-view high-magnification SEM images clearly confirm the formation of highly ordered crystalline hexagonal columns of Alq_3 MWs with a mean length and diameter of 5 μm and 1 μm as shown in Figure 2c,d, respectively.

Figure 2. Scanning electron microscopy (SEM) images of (**a**) Alq_3 NPs and (**b**) crystalline hexagonal columns of Alq_3. High-magnification SEM images of (**c**) side view and (**d**) top-view of crystal hexagonal columns of Alq_3.

3.2. Optical Properties

Figure 3a,b show the CCD images used for direct observation of the luminescence color of the Alq_3 NPs and crystal hexagonal columns of Alq_3 MWs, respectively. The color CCD image of the crystal hexagonal column of Alq_3 MWs showed much brighter luminescence than that of the Alq_3 NPs. A weak green emission was observed for the Alq_3 NPs because of their amorphous nature, as shown in Figure 3a. The crystal hexagonal column of Alq_3 MWs emitted bright green, as shown in Figure 3b. Figure 3c shows the two-dimensional (2D) LCM PL images for an isolated Alq_3 NP and crystal hexagonal column of Alq_3 MW, respectively. Under identical LCM PL imaging conditions (i.e., simultaneous comparison), a much brighter emission was observed for the crystal hexagonal column of Alq_3 MWs than for the Alq_3 single NP. The measured voltages of the photomultiplier tube output from the z-axis representing the PL intensity of the single strand of the crystal hexagonal column of Alq_3 MWs was 25–31 V, which is about 150–200 times higher than that (0.12–0.15 V) of the amorphous Alq_3 single NP, as shown in Figure 3c.

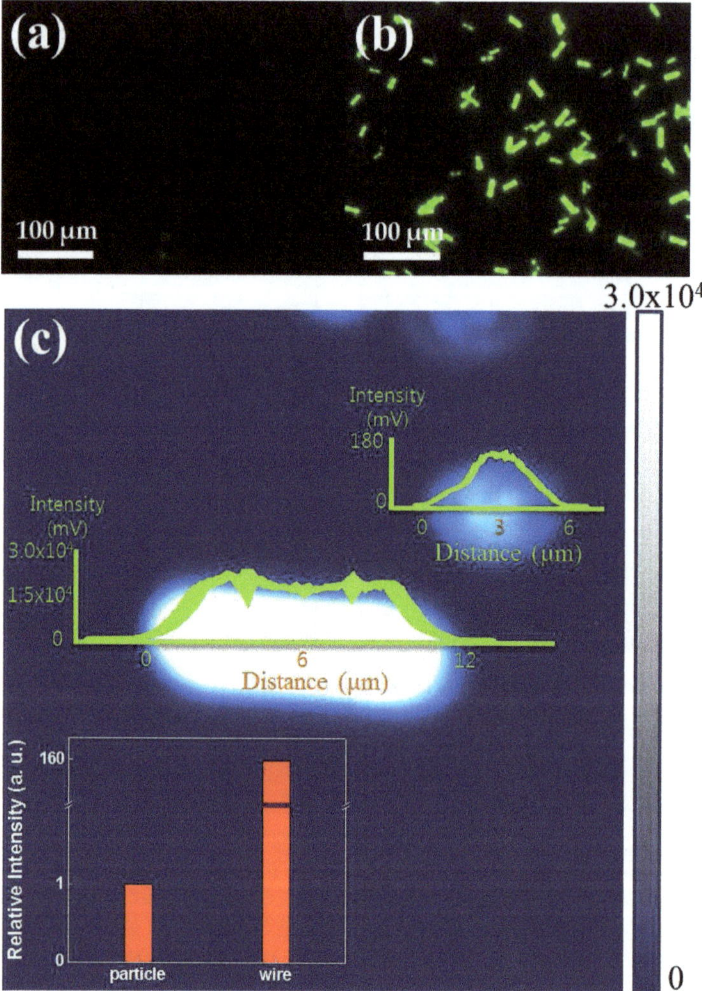

Figure 3. Color charge-couple device (CCD) images of (**a**) Alq$_3$ NPs and (**b**) crystal hexagonal column of Alq$_3$ MWs. (**c**) Photoluminescence image of both crystal Alq$_3$ MW and amorphous Alq$_3$ particle. (inset shows relative luminescence intensity of particle and wire).

Figure 4 shows the variation in the PL spectra of Alq$_3$ depending on its crystallinity and structure. A homemade laser confocal microscope system was used for nanoscale measurement of the PL for a single unit. Solution-state PL spectra were also acquired using a polymer cuvette. For the solvated crystalline Alq$_3$, the PL intensity was 1.6 times higher than that of amorphous Alq$_3$ due to auxiliary interactions between the incident light and solution, such as scattering, diffraction and reflectance. On the other hand, for nanoscale solid-state Alq$_3$ immobilized on a substrate, the individual crystals were separated and there were no auxiliary interactions. For all spectra, the main peak was positioned around 530 nm. However, the LCM PL intensity for the single unit of Alq$_3$ MWs was significantly higher due to the crystalline structure. The PL signal for a single unit of the crystal hexagonal column of Alq$_3$ MWs was 159 times more intense than that of amorphous Alq$_3$. The highly enhanced luminescence characteristics of the crystalline Alq$_3$ materials originates from the strong π-π interactions of the Alq$_3$ molecules, given that energy transfer is efficient in the Alq$_3$ crystals and there is less defect emission.

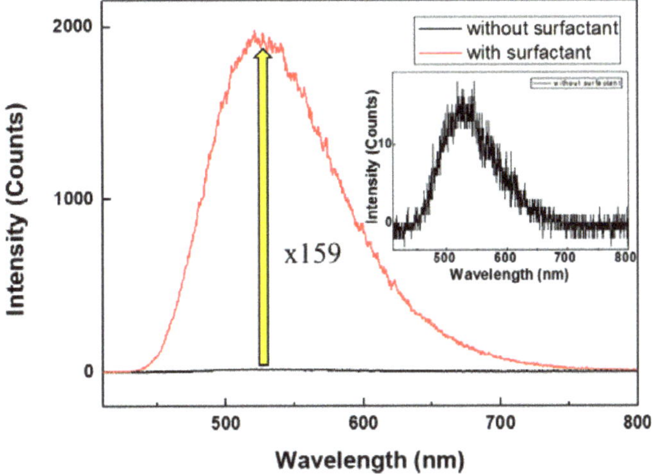

Figure 4. Comparison of solid-state laser confocal microscope (LCM) photoluminescence (PL) spectra of nanoscale crystal and amorphous Alq$_3$ (inset shows solid-state spectrum for amorphous Alq$_3$).

3.3. Structural Properties

To confirm the relationship between the crystallinity and PL efficiency of Alq$_3$, X-ray diffraction data were obtained for the samples prepared, with or without SDS as a surfactant. Figure 5 displays the XRD patterns for the two types of Alq$_3$. The XRD peaks indicate the degree of crystallinity according to Bragg's law. The peak corresponding to the (001) crystal lattice plane was observed at 6.47° for crystalline Alq$_3$. Peaks were also observed at 11.55° and 17.45° for both cases, respectively, corresponding to the (011) and (021) planes. However, other major peaks at 7.1°, 7.38°—related to the (010) and (01$\bar{1}$) planes respectively—were only observed for the crystalline structure. We can also identify the crystal phase of Alq$_3$ using XRD data. In Figure 5, the XRD pattern of Alq$_3$ MRs showed 6.47°, 7.1° and 11.55° as related to typical α-phase peaks, in addition to δ-phase peaks at 6.78° and 7.38° [23,24]. The XRD results indicate that strong π-π interactions are operative along the *b*-axis (*b* = 6.47 Å) of the Alq$_3$ molecules in the hexagonal column of Alq$_3$ MWs, which contributes to the very bright emission [25,26]. However, relatively weak XRD peaks were observed for the Alq$_3$ NPs, which have the same molar concentration as Alq$_3$ MWs—as shown in Figure 5—indicating the poor crystalline structure of the Alq$_3$ NPs. This led to a weak and broad PL spectrum for the Alq$_3$ NPs. The Alq$_3$ crystals prepared with SDS grew preferentially in a specific direction and were more crystalline than the congener prepared without SDS, based on the data shown in Figure 5.

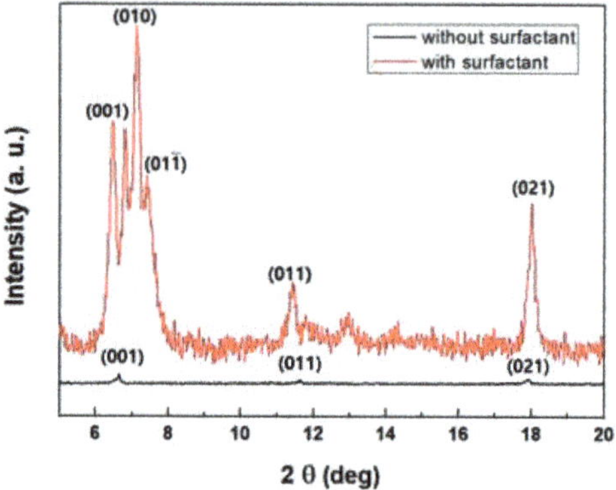

Figure 5. Comparison of X-Ray Diffraction (XRD) spectra of Alq$_3$ (upper line corresponds to crystal hexagonal column microwires (MWs), lower line indicates amorphous nanoparticles (NPs)).

4. Conclusions

Single-crystalline hexagonal columns of Alq$_3$ MWs were fabricated through a surfactant-assisted reprecipitation method. The crystalline hexagonal Alq$_3$ single MWs exhibited very bright emission with sharp peaks, as observed in the color CCD images and LCM PL images and spectra. The main PL peak at 530 nm for the crystalline Alq$_3$ single MW had a relatively narrow full-width-at-half-maximum. The intensity of the LCM PL peak at 530 nm for the single hexagonal column of Alq$_3$ MWs was approximately 160 times higher than that of the Alq$_3$ NPs, representing a dramatic increase. The highly enhanced luminescence characteristics of the crystalline Alq$_3$ materials originates from the strong π-π interactions of the Alq$_3$ molecules along the *b*-axis in the single crystalline form. Also, the surfactant present in crystals of π-π stacking has an effect on the active enhanced fluorescence and reduces the vibrational losses of the molecular structure to achieve strong fluorescence emission.

Acknowledgments: This work was supported by INHA UNIVERSITY Research Grant 55774.

Author Contributions: Seokho Kim and Dong Hyuk Park conceived and designed the experiments; Hojin Lee, Do Hyoung Kim, Jungwoon Park, Jinho Choi and Seokho Kim performed the experiments; Sun-young Kim, Seokho Kim and Dong Hyuk Park analyzed the data; Seokho Kim and Dong Hyuk Park wrote the paper.

Conflicts of Interest: The authors declare no conflict of interest.

References

1. Park, D.H.; Kim, M.S.; Joo, J. Hybrid nanostructures using π-conjugated polymers and nanoscale metals: Synthesis, characteristics, and optoelectronic applications. *Chem. Soc. Rev.* **2010**, *39*, 2439. [CrossRef] [PubMed]

2. Heeger, A.J. Semiconducting polymers: The Third Generation. *Chem. Soc. Rev.* **2010**, *39*, 2354. [CrossRef] [PubMed]

3. Wang, Q.; Ma, D. Management of charges and excitons for high-performance white organic light-emitting diodes. *Chem. Soc. Rev.* **2010**, *39*, 2387. [CrossRef] [PubMed]

4. Brooks, J.S. Organic crystals: Properties, devices, functionalization and bridges to bio-molecules. *Chem. Soc. Rev.* **2010**. [CrossRef] [PubMed]

5. Zhang, W.; Zhao, Y.S. Organic nanophotonic materials: The relationship between excited-state processes and photonic performances. *Chem. Commun.* **2016**, *52*, 8906–8917. [CrossRef] [PubMed]

6. Kim, B.H.; Kim, D.C.; Jang, M.H.; Baek, J.; Park, D.; Kang, I.S.; Park, Y.C.; Ahn, S.; Cho, Y.H.; Kim, J.; et al. Extraordinary Strong Fluorescence Evolution in Phosphor on Graphene. *Adv. Mater.* **2016**, *28*, 1657–1662. [CrossRef] [PubMed]

7. Kojima, A.; Teshima, K.; Shirai, Y.; Miyasaka, T. Organometal halide perovskites as visible-light sensitizers for photovoltaic cells. *J. Am. Chem. Soc.* **2009**, *131*, 6050–6051. [CrossRef] [PubMed]

8. Tang, C.W.; Vanslyke, S.A. Organic electroluminescent diodes. *Appl. Phys. Lett.* **1987**, *51*, 913–915. [CrossRef]

9. D'Andrade, B.W.; Forrest, S.R. White organic light-emitting devices for solid-state lighting. *Adv. Mater.* **2004**, *16*, 1585–1595. [CrossRef]

10. Huang, L.; Liao, Q.; Shi, Q.; Fu, H.; Ma, J.; Yao, J. Rubrene micro-crystals from solution routes: Their crystallography, morphology and optical properties. *J. Mater. Chem.* **2010**, *20*, 159–166. [CrossRef]

11. Garreau, A.; Duvail, J.L. Recent advances in optically active polymer-based nanowires and nanotubes. *Adv. Opt. Mater.* **2014**, *2*, 1122–1140. [CrossRef]

12. Lee, T.; Lin, M.S. Sublimation point depression of tris(8-hydroxyquinoline)aluminum(III) (Alq3) by crystal engineering. *Cryst. Growth Des.* **2007**, *7*, 1803–1810. [CrossRef]

13. Xie, W.; Fan, J.; Song, H.; Jiang, F.; Yuan, H.; Wei, Z.; Ji, Z.; Pang, Z.; Han, S. Controllable synthesis of rice-shape Alq3 nanoparticles with single crystal structure. *Phys. E Low-Dimens. Syst. Nanostruct.* **2016**, *84*, 519–523. [CrossRef]

14. Cölle, M.; Brütting, W. Thermal, structural and photophysical properties of the organic semiconductor Alq3. *Phys. Status Solidi* **2004**, *201*, 1095–1115. [CrossRef]

15. Cölle, M.; Dinnebier, R.E.; Brütting, W. The structure of the blue luminescent delta-phase of tris(8-hydroxyquinoline)aluminium(III) (Alq3). *Chem. Commun.* **2002**, *49*, 2908–2909. [CrossRef]

16. Rajeswaran, M.; Blanton, T.N. Single-crystal structure determination of a new polymorph (ε-Alq3) of the electroluminescence OLED (organic light-emitting diode) material, tris(8-hydroxyquinoline)aluminum (Alq3). *J. Chem. Crystallogr.* **2005**, *35*, 71–76. [CrossRef]

17. Bi, H.; Zhang, H.; Zhang, Y.; Gao, H.; Su, Z.; Wang, Y. Fac-Alq3 and Mer-Alq3 nano/microcrystals with different emission and charge-transporting properties. *Adv. Mater.* **2010**, *22*, 1631–1634. [CrossRef] [PubMed]

18. Cui, C.; Park, D.H.; Kim, J.; Joo, J.; Ahn, D.J. Oligonucleotide assisted light-emitting Alq3 microrods: Energy transfer effect with fluorescent dyes. *Chem. Commun.* **2013**, *49*, 5360. [CrossRef] [PubMed]

19. Park, D.H.; Kim, H.S.; Jeong, M.Y.; Lee, Y.B.; Kim, H.J.; Kim, D.C.; Kim, J.; Joo, J. Significantly enhanced photoluminescence of doped polymer-metal hybrid nanotubes. *Adv. Funct. Mater.* **2008**, *18*, 2526–2534. [CrossRef]

20. Joo, J.; Park, D.H.; Jeong, M.Y.; Lee, Y.B.; Kim, H.S.; Choi, W.J.; Park, Q.H.; Kim, H.J.; Kim, D.C.; Kim, J. Bright light emission of a single polythiophene nanotube strand with a nanometer-scale metal coating. *Adv. Mater.* **2007**, *19*, 2824–2829. [CrossRef]

21. Park, D.H.; Kim, N.; Cui, C.; Hong, Y.K.; Kim, M.S.; Yang, D.-H.; Kim, D.-C.; Lee, H.; Kim, J.; Ahn, D.J.; et al. DNA detection using a light-emitting polymer single nanowire. *Chem. Commun.* **2011**, *47*, 7944–7946. [CrossRef] [PubMed]

22. Park, D.H.; Kim, B.H.; Jang, M.K.; Bae, K.Y.; Lee, S.J.; Joo, J. Synthesis and characterization of polythiophene and poly (3-methylthiophene) nanotubes and nanowires. *Synth. Met.* **2005**, *153*, 341–344. [CrossRef]

23. Brinkmann, M.; Gadret, G.; Muccini, M.; Taliani, C.; Masciocchi, N.; Sironi, A. Correlation between Molecular Packing and Optical Properties in Different Crystalline Polymorphs and Amorphous Thin Films of mer-Tris(8-hydroxyquinoline)aluminum(III). *J. Am. Chem. Soc.* **2000**, *122*, 5147–5157. [CrossRef]

24. Rajeswaran, M.; Blanton, T.N.; Tang, C.W.; Lenhart, W.C.; Switalski, S.C.; Giesen, D.J.; Antalek, B.J.; Pawlik, T.D.; Kondakov, D.Y.; Zumbulyadis, N.; et al. Structural, thermal, and spectral characterization of the different crystalline forms of Alq3, tris(quinolin-8-olato)aluminum(III), an electroluminescent material in OLED technology. *Polyhedron* **2009**, *28*, 835–843. [CrossRef]

25. Back, S.H.; Park, J.H.; Cui, C.; Ahn, D.J. Bio-recognitive photonics of a DNA-guided organic semiconductor. *Nat. Commun.* **2016**, *7*. [CrossRef] [PubMed]

26. Park, D.H.; Jo, S.G.; Hong, Y.K.; Cui, C.; Lee, H.; Ahn, D.J.; Kim, J.; Joo, J. Highly bright and sharp light emission of a single nanoparticle of crystalline rubrene. *J. Mater. Chem.* **2011**, *21*, 8002. [CrossRef]

© 2018 by the authors. Licensee MDPI, Basel, Switzerland. This article is an open access article distributed under the terms and conditions of the Creative Commons Attribution (CC BY) license (http://creativecommons.org/licenses/by/4.0/).

Article

Understanding the Influence of Interface Morphology on the Performance of Perovskite Solar Cells

Manuel Salado [1,†] , Laura Calió [2,†] , Lidia Contreras-Bernal [3] , Jesus Idígoras [3] , Juan Antonio Anta [3] , Shahzada Ahmad [1,2,4] and Samrana Kazim [1,2,*]

[1] BCMaterials, Basque Center for Materials, Applications and Nanostructures, Bld. Martina Casiano, UPV/EHU Science Park, Barrio Sarriena, s/n, 48940 Leioa, Spain; manuel.salado@bcmaterials.net (M.S.); shahzada.ahmad@bcmaterials.net (S.A.)

[2] Abengoa Research, Abengoa, c/Energía Solar no. 1, Campus Palmas Altas, 41014 Sevilla, Spain; calio.laura@gmail.com

[3] Area de Química Física, Universidad Pablo de Olavide, E-41013 Sevilla, Spain; lconber@upo.es (L.C.-B.); jaidileo@upo.es (J.I.); jaantmon@upo.es (J.A.A.)

[4] IKERBASQUE, Basque Foundation for Science, 48013 Bilbao, Spain

* Correspondence: samrana.kazim@bcmaterials.net

† These authors contributed equally to this work.

Received: 23 May 2018; Accepted: 17 June 2018; Published: 25 June 2018

Abstract: In recent years, organo-halide perovskite solar cells have garnered a surge of interest due to their high performance and low-cost fabrication processing. Owing to the multilayer architecture of perovskite solar cells, interface not only has a pivotal role to play in performance, but also influences long-term stability. Here we have employed diverse morphologies of electron selective layer (ESL) to elucidate charge extraction behavior in perovskite solar cells. The TiO_2 mesoporous structure (three-dimensional) having varied thickness, and nanocolumns (1-dimensional) with tunable length were employed. We found that a TiO_2 electron selective layer with thickness of about c.a. 100 nm, irrespective of its microstructure, was optimal for efficient charge extraction. Furthermore, by employing impedance spectroscopy at different excitation wavelengths, we studied the nature of recombination and its dependence on the charge generation profile, and results showed that, irrespective of the wavelength region, the fresh devices do not possess any preferential recombination site, and recombination process is governed by the bulk of the perovskite layer. Moreover, depending on the type of ESL, a different recombination mechanism was observed that influences the final behavior of the devices.

Keywords: electron transport material; titanium oxide; charge dynamics; metal-halides perovskites

---

1. Introduction

During recent years, organic–inorganic halide perovskite solar cells (PSCs) have made colossal progress, competing head-to-head at lab scale with mature Silicon-based or other thin-film-based photovoltaic technologies [1]. The rapid increase in terms of power conversion efficiency (PCE) in just a few years [2–4] has been possible due to the exploration of rational charge selective contacts, interface optimization and the compositional engineering of perovskites [5,6]. The attractiveness of exploiting organic–inorganic halide perovskite is mainly due to its broad and high absorption co-efficient, along with its ambipolar character to transport electrons and holes [7]. Although the current record PCE values in PSCs have reached 22.7% [8], a thorough understanding of the processes involved in charge transport and recombination between layers and inside the perovskite layer is paramount.

Generally, in the case of photovoltaics, the incident photons create free carriers when they are absorbed by the semiconducting material. In the case of PSCs, these free carriers (electrons and holes)

travel through the perovskite to the interface of respective charge selective contacts. Depending on the deposition techniques [9–12] and the composition of the perovskite [6,13,14] used, energetic disorder may exist mainly in the grain boundaries, which act as recombination points. In order to minimize these recombination sites, extensive effort has been made to improve the quality of the perovskite layers through compositional engineering [6], interface passivation [15,16], or processing methods such as gas-phase [17,18] or vacuum assisted deposition [19,20]. However, interfaces are vulnerable to losses due to energy-level mismatch, un-intimate contact or interfacial defects which lead to high charge accumulation, which causes a drop in the performance of the solar cell, especially in open-circuit conditions [21,22]. Thus, it is vital to understand and control charge extraction across the interfaces in order to minimize energy loss and improve device performance.

Among the diverse device structures [23] investigated to date—planar [24], mesoporous [25] and inverted [26]—mesoporous-structure-based devices employing TiO_2 [27] and Spiro-OMeTAD [28] as the selective contacts for electrons and holes extraction respectively, yielded the best PCE and stability. Considerable research efforts have been focused on the hole selective layer (HSL), not only due to the high production cost of Spiro-OMeTAD, but also due to UV-instability and other disadvantages such as the need for dopants, which subsequently affects the performance of PSCs.

TiO_2 is the most ubiquitous electron selective contact in mesoporous-structure-based devices. In order to improve charge extraction and reduce nanoparticle clustering, different strategies have been implemented in the past few years. For instance, Grätzel and Ahmad et al. employed Y-TiO_2 film with the aim of improving the conductivity of TiO_2 [29]. The results showed an increment in PCE, mostly due to better charge extraction after modifying the TiO_2 layer. Following a similar strategy, Cojocaru et al. [30] carried out a post-treatment to the blocking TiO_2 layer with $TiCl_4$ and demonstrated an efficiency enhancement of 3% when comparing reverse- versus forward-measurement directions. In a similar fashion, Grätzel et al. [31] and Friend et al. [32] demonstrated beneficial effects when mesoporous TiO_2 is post-treated with Li^{2+}. It was argued that lithium salt can passivate the interface and can reduce the trap density. These effects led to the fabrication of devices with not only reduced hysteresis and favored electron extraction, but also improved stability.

In the recent past, electrochemical analysis such as impedance spectroscopy (IS), Intensity-modulated photocurrent spectroscopy (IMPS) and intensity-modulated photovoltage spectroscopy (IMVS) are being employed to elucidate charge recombination, accumulation and transport [33,34]. Different approaches have been proposed, from the typical perovskite device architecture where the electron selective layer (ESL) is either high temperature processed mesoporous TiO_2, or low temperature processed TiO_2 in a simple planar structures that are ideal for flexible configurations. Previously, we reported [35] how the use of TiO_2 nanocolumns affects the charge dynamics on devices, along with improvement to stability.

Herein, we report the role of TiO_2 microstructure (3-dimensional and 1-dimensional) and thickness on charge dynamic studies in polycrystalline layers of methyl ammonium lead triiodide (MAPbI$_3$). The optimization of the ESL thickness was studied by IS, to optimize charge separation and elucidate its role on charge dynamics, suggesting that, irrespective of its microstructure, an electron extraction layer with a thickness of ca. 100 nm produces better device performance in our experimental conditions.

2. Materials and Methods

2.1. Materials

All chemicals were commercial products and procured from either Sigma-Aldrich (St. Louis, MO, USA) or Acros organics (Geel, Belgium) and were employed as such. 2,2′,7,7′-tetrakis(N,N-di-p-methoxyphenyamine)-9,9-spirobifluorene (Spiro-OMeTAD) was acquired from Merck KGaA (Darmstadt, Germany). Methylamine iodide, CH_3NH_3I (MAI) was bought from Greatcells Solar (Queanbeyan, New South Wales, Australia) and PbI_2 were bought from TCI (Tokyo, Japan). Chlorobenzene and anhydrous DMSO were obtained from Sigma-Aldrich (St. Louis, MO, USA). Acetone, acetonitrile solvents were bought from Acros organics.

2.2. Electron Selective Layer (ESL) Deposition

Vertically aligned nanocolumns with a thickness of 100 or 200 nm were prepared at room temperature by physical vapor deposition at oblique incidence (GLAD-PVD) in an electron beam evaporator (Advanced Products and Technologies GmbH, Nurtingen, Germany), as described in previous publications [35,36]. The deposition time was controlled to achieve two different nanostructured thicknesses, i.e., 100 or 200 nm. In case of TiO_2 mesoporous layer (30NRD, Greatcells Solar, Queanbeyan, Australia), to obtain different thicknesses, mesoporous stock solutions with different dilution (1:7 and 1:3.5 *w/w*) were prepared and spun-coated at 4000 rpm for 20 s. For the fabrication of planar devices, no such layer was used. All the ESL-based substrates were then annealed up to 450 °C in a progressive manner for 2 h.

2.3. Device Preparation

To fabricate the PSCs, laser-etched FTO-coated glass (TEC15, Pilkington, Tokyo, Japan) was used. Prior to use, the electrodes were cleaned using Hellmanex solution and washed with deionized water. Subsequently, they were ultrasonicated for 25 min at 60 °C in acetone, 2-propanol, and finally dried using nitrogen blow. The compact layer (hole-blocking layer) was made up of TiO_2, deposited by spray pyrolysis at 450 °C by diluting 1 mL of titanium diisopropoxide bis(acetyl acetonate) precursor solution (75% in 2-propanol, Sigma-Aldrich) in 19 mL of pure ethanol and dry air as carrier gas. These substrates were then kept for a further 30 min at 450 °C. Once the TiO_2 deposited electrodes attained room temperature, they were immersed in a 0.02 M $TiCl_4$ solution in deionized water at 70 °C for 30 min to form a homogeneous layer. After this treatment, the electrodes were washed by using deionized water and annealed at 500 °C for 30 min, and then allowed to cool down to room temperature.

For deposition of the perovskite layer, 1.2 M of PbI_2 and MAI precursor solutions (1:1 ratio), in DMSO were prepared inside an argon-filled glovebox (H_2O level: <1 ppm and O_2 level: <10 ppm) and left for stirring overnight at 80 °C. The perovskite layer was deposited using the so called one-step method followed by solvent engineering; for this the precursor solution was spun-coated on top of the mesoporous layer at 1000 rpm for 10 s and then 6000 rpm for 20 s. Chlorobenzene was dripped during the second stage of spinning process. Following this the electrodes were transferred to a hotplate and annealed at 100 °C for 1 h to allow the formation of perovskites. The hole transport layer was then deposited on top of the perovskites; for this 35 μL of a Spiro-OMeTAD solution was then spun-coated at 4000 rpm for 20 s. Spiro-OMeTAD material (70 mM) was dissolved in 1 mL of chlorobenzene and standard additives such as 17.5 μL of a lithium bis(trifluoromethylsulphonyl)imide (LiTFSI) stock solution (520 mg of LiTFSI in 1 mL of acetonitrile), 21.9 μL of a FK209 (Tris(2-(1H-pyrazol-1-yl)-4-tert-butylpyridine) cobalt(III)Tris(bis(trifluoromethylsulfonyl)imide))) stock solution (400 mg in 1 mL of acetonitrile), and 28.8 μL of 4-tert-butylpyridine (*t*-BP). The device was finished by depositing 80 nm of gold as a cathode layer, using thermal evaporation under a vacuum level between 10^{-6} to 10^{-5} torr.

For IS measurements, light sources were provided using red (λ = 635 nm) and blue (λ = 465 nm) light emitting diodes (LEDs) in a broad range of DC light intensities. By doing so, the devices could be probed on two different optical penetrations [37]. To record the impedance spectra, a response analyzer module ((PGSTAT302N/FRA2, Metrohm Autolab, Utrecht, Netherlands) was employed, and 20 mV perturbation in the 10^7–10^{-2} Hz range was applied for the measurements. The voltage drop (arising due to series resistance) was avoided by performing the measurements at the open-circuit potential, and the Fermi level (linked to the open-circuit voltage) was fixed by the DC (bias) illumination intensity. The different response under blue and red light was compensated by monitoring all parameters and was plotted as a function of the open-circuit potential generated by individually bias light.

3. Results and Discussion

Figure 1a represents a schematic of the fabricated solar cell, in which the different layers (FTO/bl-TiO$_2$/mesoporous or nanocolumns TiO$_2$/CH$_3$NH$_3$PbI$_3$/Spiro-OMeTAD/Au) can be visualized. In order to observe morphological difference of the interface between the absorber layer and the electron extraction layer, scanning electron microscopy (SEM, Hitachi, Tokyo, Japan) experiments were performed. Figure 1b,c represents the cross-sectional SEM images of the fabricated devices having different thicknesses of mesoporous structure. The thickness of the mesoporous structure was tuned by varying the dilution rate of the TiO$_2$ paste; for example, the more diluted TiO$_2$ paste led to a thin layer (Figure 1 and Table S1). In case of TiO$_2$ nanocolumns, the length of columns was varied by increasing the deposition time. Considering the values in Table S1, the total thickness of photoactive layer (mesoporous & perovskite capping layer) was c.a. 300 nm. It can be seen from the figures that the thickness of the perovskite capping layer was also altered when the dilution rate was modified, having a strong influence in the series resistance of the device (Figure S1). The aim of using different microstructures (3D and 1D) was to optimize the effective thickness for ideal perovskite infiltration and electron extraction. According to the current density-voltage (*J-V*) curves (Figure 2), the best configuration was found to be 1:7, which allowed a thickness of around 100 nm (Table S1).

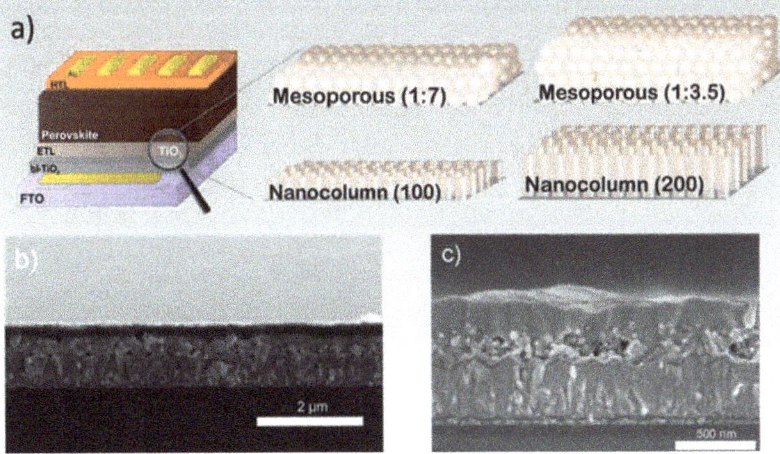

Figure 1. (**a**) Cross-sectional device scheme and electron selective layer (ESL) used, (**b**) scanning electron microscopy (SEM) image of PSC with 1:7 TiO$_2$ mesoporous structure and (**c**) SEM image of PSC with 1:3.5 TiO$_2$ mesoporous structure.

TiO$_2$ nanostructure and its thickness can influence the properties of the layer deposited atop of it. Earlier, we have presented the cross-sectional image of 1-D TiO$_2$ nanocolumns (100 and 200 nm) [35]. To study the role of TiO$_2$ microstructure effects on the band gap of the perovskite, we calculated the optical band gap of devices having different ESL configurations using the Tauc plot method. The absorption coefficient (α) was calculated from measured absorbance spectra (Figure S2a) and a Tauc plot $(\alpha h\nu)^2$ vs. $h\nu$ was used to calculate the optical direct band gap as shown in Figure S2b. The linear extrapolation of this curve to intercept the horizontal $h\nu$ axis gives the value of band gap (~1.59 eV).

Figure 2 illustrates the *J-V* curves of fabricated PSCs using different ESL under standard conditions (100 mW·cm^{-2}—AM 1.5 illumination) at room temperature. Table 1 summarizes the different photovoltaic parameters of the best-performing devices for each configuration, while their statistical data are listed in Table S2. As expected, the thickness of the ESL was found to influence the photovoltaic performance.

A lower value of V_{oc} was noted (V_{oc} = 930 mV) with a thicker TiO_2 layer (nanocolumns-200 nm and nanoparticles-1:3.5), in comparison to the thinner TiO_2 ESLs (nanocolumns-100 nm and nanoparticles-1:7), or planar devices (0 nm). Although no significant changes were observed in terms of the J_{sc}, and regardless of the nanostructure and thickness of the ESLs, all devices gave J_{sc} in the range of 18–19 mA·cm^{-2}. These results are in line with the absorbance spectra obtained for the different configuration (Figure S2). Devices fabricated with mesoporous layer (1:7, Ø30 nm) and 100 nm nanocolumns yielded best PCE, as well as lower series resistance (Figure S1).

In PSCs, the series resistance (R_s) is known as a crucial factor which can affects the device fill factor, and the R_s depends not only on bulk resistance of the photoactive layer, but also on metallic contacts and contact resistance at each interface. The R_s of the devices (obtained from the software by fitting the J-V curve) based on 100 nm TiO_2 nanocolumns and thin mesoporous layers (1:7 dilutions, thickness around 125 nm) show a lower R_s value than the 200 nm nanocolumns and thick mesoporous layer (1:3.5). The reduced value of R_s agrees with the enhanced FF for thin TiO_2 ESLs (mp-1:7 and 100 nm NC), irrespective of their morphology.

Dissimilarities in device behavior can be also analyzed from the dark-current measurements (represented by the hollow symbols in Figure 2). It is known that the V_{oc} is strongly related to the onset voltage point of the device measured in the dark under a forward bias, and delay in the onset voltage point of the device will result in high V_{oc}. Devices prepared with mesoporous TiO_2 layer (1:7, Ø30 nm), 100 nm TiO_2 nanocolumns and planar devices exhibit comparatively lower dark current and late switch-on of the devices under forward bias than the thicker mesoporous (1:3.5, Ø30 nm), and 200 nm nanocolumns. The delay in the onset of the dark current for the abovementioned devices is well supported, with comparatively high values of V_{oc} (Table 1), indicating reduced charge recombination rates in thinner ESLs, which will be further demonstrated by impedance measurements in the next section.

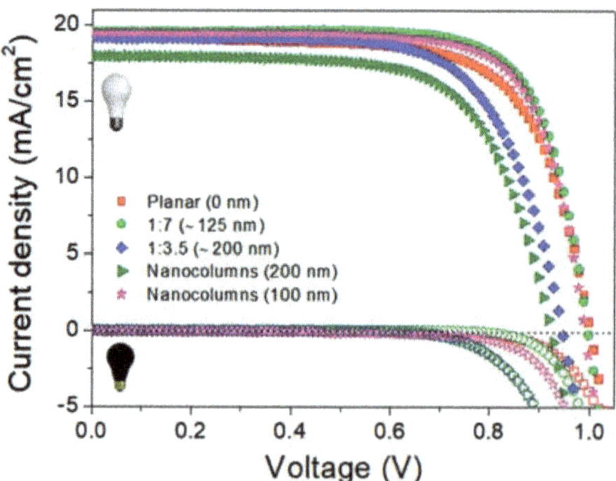

Figure 2. Current density-voltage (J-V) characteristics of the best performing PSCs measured in dark and under illumination at AM 1.5G, 100 mW/cm^2 in reverse bias scans performed at 100 mV s^{-1}, for different electron selecting layers. Hollow symbols indicate the dark current of the different ESL-based devices.

However, an ESL based on non-planar structure helps to minimize the hysteresis. As shown in Table S3, all non-planar devices show a hysteresis index (HI) of around 0.1–0.2, whereas a value of 0.48 was found for planar devices.

HI values for the $MAPbI_3$ based PSCs using different ESLs were calculated (Table S3) along with the J-V parameters in forward and reverse directions for the best-performing device. Hysteresis behavior has been attributed to several causes, such as ionic migration, charge accumulation at selective contacts interfaces or distortion of the octahedral structure, or a combination of all of these factors. Recently, research direction has been focused on the slow processes occurring in the PSCs to study possible cause of hysteresis. The charge accumulation at the TiO_2/perovskite interface coupled with ionic migration may cause this phenomenon [23,30,38].

Table 1. *J-V* characteristics parameters of best-performing device based on MAPbI$_3$ perovskite with different electron selecting layer. * The value in parenthesis depicts the thickness used.

Device Configuration *	V_{oc} (mV)	J_{sc} (mA/cm^2)	Fill Factor (%)	Efficiency (%)
Planar (0 nm)	1000	19.28	69.04	13.45
mp-1:3.5 (~200 nm)	940	19.15	68.60	12.42
mp-1:7 (~125 nm)	990	19.55	75.07	14.61
Nanocolumns (200 nm)	920	17.94	68.07	11.26
Nanocolumns (100 nm)	990	19.44	74.47	14.35

In order to unravel the charge dynamics depending on the different ESLs, IS experiments were performed. This technique provides information about various internal processes occurring in working devices in different operation conditions and offers the possibility of analyzing interfacial and bulk processes separately. The resistive and capacitive processes occurring in the device are related to bulk and interfacial processes such as charge transport, accumulation, recombination and ion-mediated processes. Charge accumulation and ion migration can be probed by IS, and both processes are possibly interconnected and contribute to the hysteresis behavior. Charge imbalance also promotes an intrinsic instability, along with an irrational interface, both of which can accelerate the degradation of the fabricated devices. In our case, we have observed that devices with higher thickness (non-planar devices) of the ESL exhibited smaller values of HI than planar devices.

Figure 3 shows the apparent capacitance versus frequency plots, illustrating the different processes depending on the range of frequency. Low-frequency processes could be related to hysteresis behavior and thus we studied the variation of the low-frequency capacitance as a function of photo-voltage (Figure S3). The value of slope (V^{-1}) close to 22 indicates the existence of an accumulation regime. Our results suggest that planar configuration presents a higher slope than non-planar-based structure, suggesting a possible higher charge accumulation at the interface.

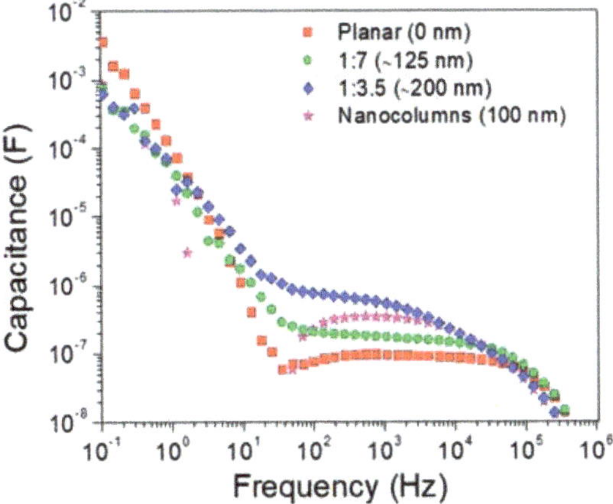

Figure 3. Frequency dependent apparent capacitance for PSCs using different electron selective layer at 0.9 V.

Furthermore, we undertook experiments to measure the charge dynamics that determine the functioning of PSCs. For this, we performed IS measurements with two different excitation wavelengths (blue and red) [37]. Based on the different penetration depths, which provoke a dissimilar charge generation profile with the use of two illumination wavelengths, the main purpose of this characterization was to detect in-homogeneities or the existence of a favored recombination region inside the solar device. By implementing this, we can study recombination processes occurring close to the electron (blue) and hole (red) in a more homogeneous way, as we will create different charge-generation profiles. As shown in Figure 4, two arcs with different characteristic frequencies can be distinguished for all the analyzed samples; one on the low frequency (LF) region (0.1–10 Hz), and one in the high-frequency (HF) region (10^2–10^5 Hz). The semicircle at high frequencies can be directly linked to the recombination kinetics in the devices, while the semicircle at low frequencies, although also indirectly affected by recombination, is related to slow processes such as ionic migration or charge accumulation at the interfaces. We have found that IS measurements under red and blue light coincide with each other (Figure 4). By fitting the impedance response under blue and red illumination for $MAPbI_3$/Spiro devices to a simple -R_S-(R_{LF},-CPE_1)-(R_{HF},-CPE_2)-equivalent circuit, the HF (recombination) resistance can be extracted (Figure 5). In this analysis, a constant phase element (CPE) replaces an ideal capacitor in order to obtain a better fitting. The resulting R_{HF} values under red and blue light excitation also coincide quite well, confirming that fresh samples do not possess any preferential recombination site. This fact points toward recombination processes occurring in the bulk of the perovskite layer, as previously reported [37].

Figure 6 shows the R_{HF} values under red light excitation for all samples studied. Different slopes were identified depending on the ESL configuration. Previous studies reported the relationship between these slopes and the predominant recombination mechanism inside the solar device (β parameter in Figure 6) [37,39]. Our results (Table S4), suggest that the planar configuration presents a different or mixed recombination mechanism ($n = 1/\beta = 1.41$) with respect to the trap-limited recombination mechanism observed in the others studied configuration ($n = 1/\beta$ ~2). Considering the consistency between the recombination resistance data obtained with red and blue optical penetration profiles, which points toward bulk recombination, the different ideality factors obtained suggest that electrical features of perovskite depend on the nature of the ESL on which it is deposited. Even though the band gap of the perovskite remains unaltered, the crystallization process might be different, which will affect the transport and recombination features inside the active perovskite layer. It is well reported that the capacitance in the high-frequency range of the PSC is mostly governed by the geometrical component, which can be expressed as following; $C_g = \varepsilon \cdot \varepsilon_0 \cdot A/d$. Here ε is the dielectric constant of the perovskite, ε_0 is the permittivity in vacuum, A is the effective contact area and d is the thickness of the absorber layer. The value of capacitance extracted from the high-frequency component (Figure S4) was constant on bias voltage; this is in accordance with its geometrical nature.

The role of ESL in devices' degradation was monitored, by recording J-V measurements and incident photon-to-electron conversion efficiencies (IPCE) after 30 days (Figure S5a,b). We found that non-planar devices were more stable than planar ones. IPCE measurements of aged devices showed a systematic depletion of the photo generation in the full-wavelength range. Furthermore, following the methodology of previous works [37,39], we also measured the open-circuit voltage (V_{oc}) as a function of temperature (T in K) in order to extract the band gap of aged devices. Figure S6 represents the open-circuit voltage versus temperature curve of aged devices (stored under humid conditions for 30 days), and fits well into the straight line. The activation energy—which gives information about dominant recombination pathway—was estimated by extrapolation of the data to T = 0 K. It can be observed that the calculated band gaps show increment with respect to fresh devices (E_g calculated by means of the absorbance measurements). These results are in accordance with the previous published results [37].

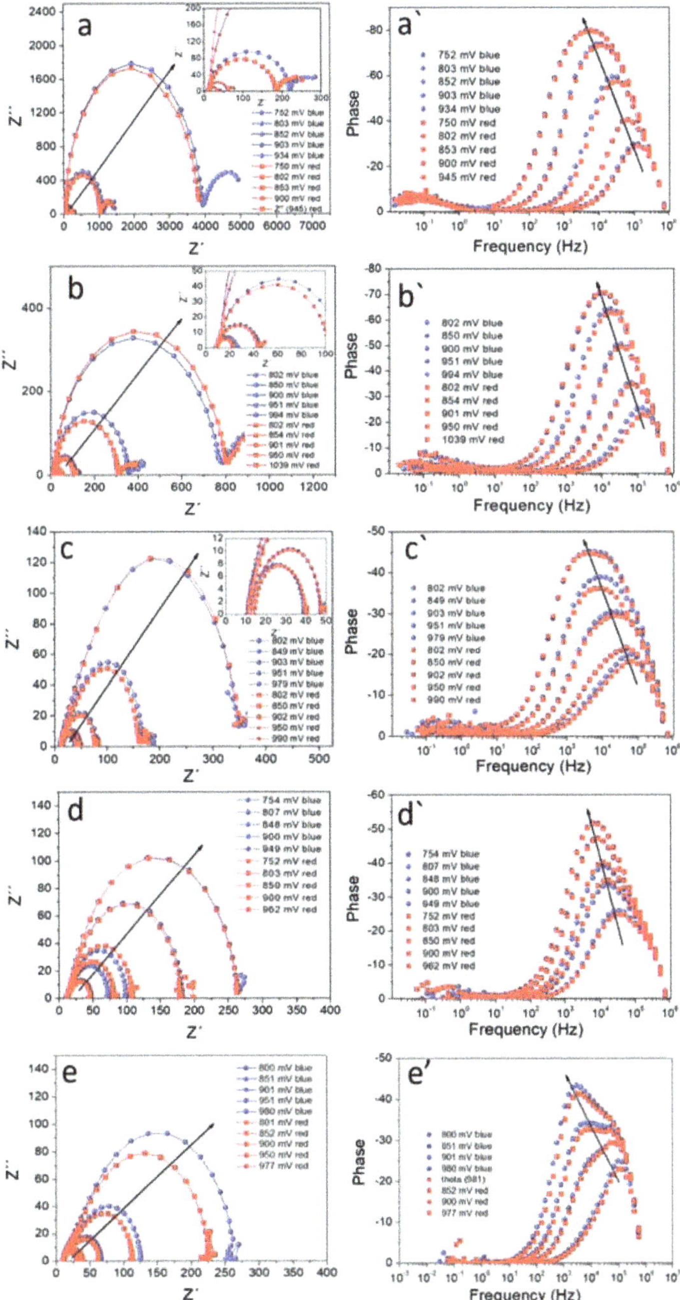

Figure 4. Nyquist (**a–e**) and Bode plots (**a′–e′**) at different applied voltages for CH$_3$NH$_3$PbI$_3$ perovskite solar cells using two excitation wavelengths, λ_{blue} = 465 nm and λ_{red} = 635 nm. Different configurations are labeled as follows: (**a,a′**) planar, (**b,b′**) mesoporous 1:7, (**c,c′**) mesoporous 1:3.5, (**d,d′**) nanocolumns 100 nm and (**e,e′**) nanocolumns 200 nm. Arrows indicate the evolution of the measurements when we measured from V_{OC} to lower values.

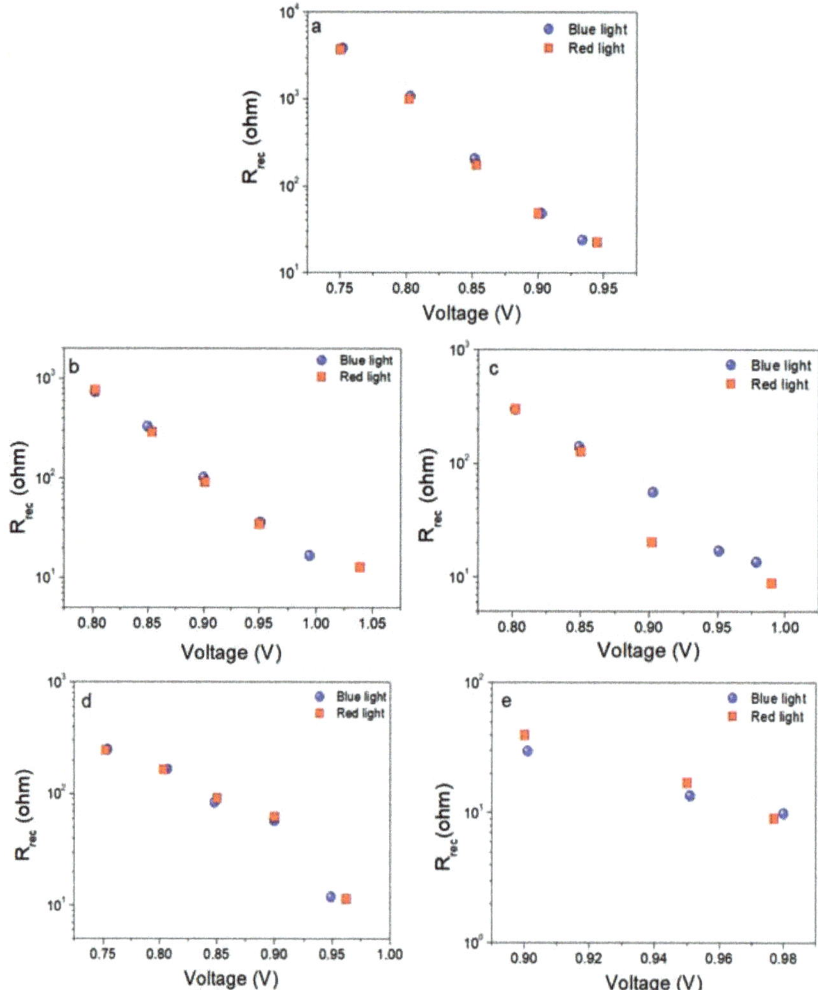

Figure 5. Recombination resistance plots (**a**–**e**) at different applied voltages for $CH_3NH_3PbI_3$ perovskite using the two excitation wavelengths, λ_{blue} = 465 nm and λ_{red} = 635 nm. Different configurations are labeled as follows: (**a**) planar, (**b**) mesoporous 1:7, (**c**) mesoporous 1:3.5, (**d**) nanocolumns 100 nm and (**e**) nanocolumns 200 nm.

We propose that the increase in the band gap can be attributed to the emergence of lead iodide crystal due to a partial degradation of the perovskite material which can be located at the TiO_2/perovskite interface. It is worth noting that nanocolumns (100 nm) and mesoporous (1:7) present the lower deviation in E_g with respect to the initial values, indicating their better stability under aging. In the case of planar devices, a decrease in the band gap was noted. The decrease in V_{oc}, can be attributed to the formation of defect sites, which further induce new recombination sites.

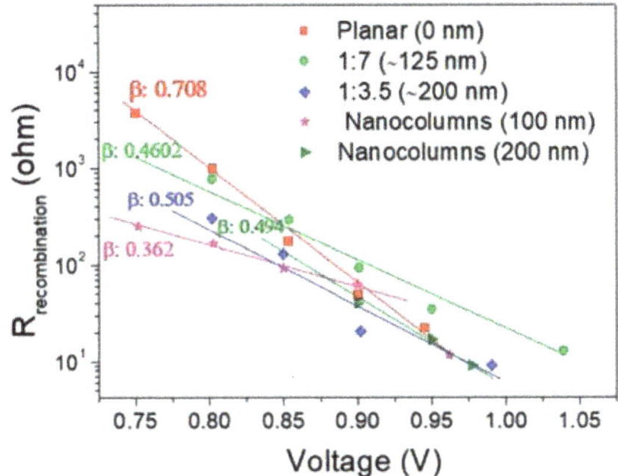

Figure 6. The representative recombination resistances ($R_{recombination}$) of the different TiO_2 structures samples determined from IS under different applied bias voltages irradiated with red light.

4. Conclusions

Interfaces at the perovskite perform a crucial role in charge separation and transport for solar cells fabrication. Electron selective contacts, having different microstructure (3-dimensional and 1-dimensional) and thickness of porous TiO_2 layer, were fabricated. The thickness and morphology of the electron selecting layer influenced the device performance and stability, possibly by affecting the perovskite crystallization process or the charge transfer at the interface. In the case of planar devices, a higher hysteresis index was observed, while both mesoporous (3D) and nanocolumns layer (1D) with 100 nm thickness gave good performance and stability.

By using impedance spectroscopy measurements, the samples were illuminated with different wavelengths (blue and red), and similar recombination resistance data were obtained irrespective of wavelength region, suggesting that the recombination process at open circuit occurs in the bulk of the perovskite layer. However, depending on the type of electron selective layer deposited, a different recombination path was observed as demonstrated by a different value of ideality factor, suggesting that electrical features of a perovskite layer depend on the nature of the ESL on which they are deposited. This work contributes to an understanding of the role of electron selective contact and its bearing on charge kinetics in perovskite solar cells.

Supplementary Materials: The following are available online at http://www.mdpi.com/1996-1944/11/7/1073/s1, Figure S1: Series resistance under 1 sun-illumination for different electron transport structures with reverse scans performed at 0.1 V s^{-1}, Figure S2: (a) Absorption spectra of $CH_3NH_3PbI_3$ perovskite infiltrated in different TiO_2 structures. (b) Tauc plot to estimate optical band gap for perovskite films in the presence of TiO_2 mesoporous or nanocolumnar structures having different thickness, Figure S3: Low frequency capacitance as a function of open circuit voltage of the PSCs using TiO_2 electron transport layer, Figure S4: Geometric capacitance at different applied voltages for $CH_3NH_3PbI_3$ perovskite using two excitation wavelengths λ_{blue} = 465 nm and λ_{red} = 635 nm, Figure S5: J-V and IPCE measured after 30 days at humid conditions; (a) dark and under white light illumination, (b) IPCE and calculated integrated short circuit current, Figure S6: Open-circuit voltage as a function of temperature for PSCs with different TiO_2 configurations under white light intensity of 14.15 W/m^2, Table S1: Layer thicknesses extracted from SEM images of $MAPbI_3$ perovskite solar cells using different thickness of TiO_2 ESL, Table S2: Statistical data of photovoltaics parameters of $MAPbI_3$-based PSCs, Table S3: J-V characteristic parameters values from the reverse and forward scan directions and calculated hysteresis index of the different ESL configurations, Table S4: Ideality factors of the perovskite solar cells acquired from impedance measurement using different electron transport configuration.

Author Contributions: M.S. planned the experiments and analyzed IS measurements and wrote the first draft of the paper. L.C. fabricated and characterized the devices. L.C.-B. performed one of the IS experiments, J.I. and J.A.A. provided guidance. S.K. coordinated and edited the draft for final document preparation, and S.A. directed and provide guidance to the research. All authors contributed to the completion of the final document.

Funding: This work was supported by FP7-Marie curie international training network [Thinface-607232] and also partially supported by a European Research Council grant, [MOLEMAT-72760]. J.A.A., J.I. thanks Junta de Andalucía for financial support via grant FQM 1851 and FQM 2310, Ministerio de Economía y Competitividad of Spain under grants MAT2013-47192-C3-3-R and MAT2016-76892-C3-1-R and Red de Excelencia "Emerging Photovoltaic Technologies".

Acknowledgments: Authors thank Manuel Oliva and Agustin Rodríguez González-Elipe for nanocolumns preparation.

Conflicts of Interest: The authors declare no conflict of interest.

References

1. Tao, M. Inorganic Photovoltaic solar cells: Silicon and beyond. *Electrochem. Soc. Interface* **2008**, *17*, 30–35.
2. Green, M.A.; Ho-Baillie, A.; Snaith, H.J. The emergence of perovskite solar cells. *Nat. Photonics* **2014**, *8*, 506–514. [CrossRef]
3. Park, N.G. Perovskite solar cells: An emerging photovoltaic technology. *Mater. Today* **2015**, *18*, 65–72. [CrossRef]
4. Kim, H.S.; Im, S.H.; Park, N. Organolead halide perovskite: New horizons in solar cell research. *J. Phys. Chem. C* **2014**, *118*, 5615–5625. [CrossRef]
5. Jeon, N.J.; Noh, J.H.; Kim, Y.C.; Yang, W.S.; Ryu, S.; Seok, S.I. Solvent engineering for high-performance inorganic-organic hybrid perovskite solar cells. *Nat. Mater.* **2014**, *13*, 897–903. [CrossRef] [PubMed]
6. Jeon, N.J.; Noh, J.H.; Yang, W.S.; Kim, Y.C.; Ryu, S.; Seo, J.; Seok, S.I. Compositional engineering of perovskite materials for high-performance solar cells. *Nature* **2015**, *517*, 476–480. [CrossRef] [PubMed]
7. Kazim, S.; Nazeeruddin, M.K.; Grätzel, M.; Ahmad, S. Perovskite as light harvester: A game changer in photovoltaics. *Angew. Chem. Int. Ed.* **2014**, *53*, 2812–2824. [CrossRef] [PubMed]
8. NREL Efficiency Chart. Available online: http://www.nrel.gov/ncpv/images/efficiency_chart.jpg (accessed on 12 January 2018).
9. Burschka, J.; Pellet, N.; Moon, S.J.; Humphry-Baker, R.; Gao, P.; Nazeeruddin, M.K.; Grätzel, M. Sequential deposition as a route to high-performance perovskite-sensitized solar cells. *Nature* **2013**, *499*, 316–319. [CrossRef] [PubMed]
10. Liu, M.; Johnston, M.B.; Snaith, H.J. Efficient planar heterojunction perovskite solar cells by vapour deposition. *Nature* **2013**, *501*, 395–398. [CrossRef] [PubMed]
11. Sutherland, B.R.; Hoogland, S.; Adachi, M.M.; Kanjanaboos, P.; Wong, C.T.O.; McDowell, J.J.; Xu, J.; Voznyy, O.; Ning, Z.; Houtepen, A.J.; et al. Perovskite thin films via atomic layer deposition. *Adv. Mater.* **2015**, *27*, 53–58. [CrossRef] [PubMed]
12. Barrows, A.T.; Pearson, A.J.; Kwak, C.K.; Dunbar, A.D.F.; Buckley, A.R.; Lidzey, D.G. Efficient planar heterojunction mixed-halide perovskite solar cells deposited via spray-deposition. *Energy Environ. Sci.* **2014**, *7*, 2944–2950. [CrossRef]
13. Pellet, N.; Gao, P.; Gregori, G.; Yang, T.Y.; Nazeeruddin, M.K.; Maier, J.; Grätzel, M. Mixed-organic-cation perovskite photovoltaics for enhanced solar-light harvesting. *Angew. Chem. Int. Ed.* **2014**, *53*, 3151–3157. [CrossRef] [PubMed]
14. Cho, K.T.; Paek, S.; Grancini, G.; Carmona, C.R.; Gao, P.; Lee, Y.H.; Nazeeruddin, M.K. Highly efficient perovskite solar cells with a compositional engineered perovskite/hole transporting material interface. *Energy Environ. Sci.* **2017**, *10*, 621–627. [CrossRef]
15. Abate, A.; Saliba, M.; Hollman, D.J.; Stranks, S.D.; Avolio, R.; Petrozza, A.; Snaith, J. Supramolecular Halogen Bond Passivation of Organometal-Halide Perovskite Solar Cells. *Nano Lett.* **2014**, *14*, 3247–3254. [CrossRef] [PubMed]
16. Hwang, I.; Jeong, I.; Lee, J.; Ko, M.J.; Yong, K. Enhancing Stability of Perovskite Solar Cells to Moisture by the Facile Hydrophobic Passivation. *ACS Appl. Mater. Interfaces* **2015**, *7*, 17330–17336. [CrossRef] [PubMed]
17. Hao, F.; Stoumpos, C.C.; Liu, Z.; Chang, R.P.H.; Kanatzidis, M.G. Controllable perovskite crystallization at a gas-solid interface for hole conductor-free solar cells with steady power conversion efficiency over 10%. *J. Am. Chem. Soc.* **2014**, *136*, 16411–16419. [CrossRef] [PubMed]
18. Bhachu, D.S.; Scanlon, D.O.; Saban, E.J.; Bronstein, H.; Parkin, I.P.; Carmalt, C.J.; Palgrave, R. Scalable Route to $CH_3NH_3PbI_3$ Perovskite Thin Films by Aerosol Assisted Chemical Vapor Deposition. *J. Mater. Chem. A* **2015**, *3*, 9071–9073. [CrossRef]

19. Yin, J.; Qu, H.; Cao, J.; Tai, H.; Li, J.; Zheng, N. Vapor-assisted crystallization control toward high performance perovskite photovoltaics with over 18% efficiency in the ambient atmosphere. *J. Mater. Chem. A* **2016**, *4*, 13203–13210. [CrossRef]

20. Chen, Q.; Zhou, H.; Hong, Z.; Luo, S.; Duan, H.-S.; Wang, H.-H.; Liu, Y.; Li, G.; Yang, Y. Planar Heterojunction Perovskite Solar Cells via Vapor-Assisted Solution Process. *J. Am. Chem. Soc.* **2013**, *136*, 622–625. [CrossRef] [PubMed]

21. Kirchartz, T.; Staub, F.; Rau, U. Impact of Photon-Recycling on the Open-Circuit Voltage of Metal Halide Perovskite Solar Cells. *ACS Energy Lett.* **2016**, *1*, 731–739. [CrossRef]

22. Yan, W.; Li, Y.; Li, Y.; Ye, S.; Liu, Z.; Wang, S.; Bian, Z.; Huang, C. High-performance hybrid perovskite solar cells with open circuit voltage dependence on hole-transporting materials. *Nano Energy* **2015**, *16*, 428–437. [CrossRef]

23. Wang, Y.; Wang, H.Y.; Han, J.; Yu, M.; Hao, M.Y.; Qin, Y.; Fu, L.M.; Zhang, J.P.; Ai, X.C. The Influence of Structural Configuration on Charge Accumulation, Transport, Recombination, and Hysteresis in Perovskite Solar Cells. *Energy Technol.* **2017**, *5*, 442–451. [CrossRef]

24. Baena, J.P.C.; Steier, L.; Tress, W.; Saliba, M.; Neutzner, S.; Matsui, T.; Giordano, F.; Jacobsson, T.J.; Kandada, A.R.S.; Zakeeruddin, S.M.; et al. Highly efficient planar perovskite solar cells through band alignment engineering. *Energy Environ. Sci.* **2015**, *8*, 2928–2934. [CrossRef]

25. Choi, J.J.; Yang, X.; Norman, Z.M.; Billinge, S.J.L.; Owen, J.S. Structure of methylammonium lead iodide within mesoporous titanium dioxide: Active material in high-performance perovskite solar cells. *Nano Lett.* **2014**, *14*, 127–133. [CrossRef] [PubMed]

26. Meng, L.; You, J.; Guo, T.-F.; Yang, Y. Recent Advances in the Inverted Planar Structure of Perovskite Solar Cells. *Acc. Chem. Res.* **2016**, *49*, 155–165. [CrossRef] [PubMed]

27. Murugadoss, G.; Mizuta, G.; Tanaka, S.; Nishino, H.; Umeyama, T.; Imahori, H.; Ito, S. Double functions of porous TiO_2 electrodes on $CH_3NH_3PbI_3$ perovskite solar cells: Enhancement of perovskite crystal transformation and prohibition of short circuiting. *APL Mater.* **2014**, *2*, 081511. [CrossRef]

28. Li, M.-H.; Hsu, C.-W.; Shen, P.-S.; Cheng, H.-M.; Chi, Y.; Chen, P.; Guo, T.-F. Novel spiro-based hole transporting materials for efficient perovskite solar cells. *Chem. Commun.* **2015**, *51*, 15518–15521. [CrossRef] [PubMed]

29. Qin, P.; Domanski, A.L.; Chandiran, A.K.; Berger, R.; Butt, H.-J.; Dar, M.I.; Moehl, T.; Tetreault, N.; Gao, P.; Ahmad, S.; et al. Yttrium-substituted nanocrystalline TiO_2 photoanodes for perovskite based heterojunction solar cells. *Nanoscale* **2014**, *6*, 1508–1514. [CrossRef] [PubMed]

30. Cojocaru, L.; Uchida, S.; Jayaweera, P.V.V.; Kaneko, S.; Wang, H.; Nakazaki, J.; Kubo, T.; Segawa, H. Effect of TiO_2 Surface Treatment on the Current–Voltage Hysteresis of Planar-Structure Perovskite Solar Cells Prepared on Rough and Flat Fluorine-Doped Tin Oxide Substrates. *Energy Technol.* **2017**, *5*, 1762–1766. [CrossRef]

31. Giordano, F.; Abate, A.; Baena, J.P.C.; Saliba, M.; Matsui, T.; Im, S.H.; Zakeeruddin, S.M.; Nazeeruddin, M.K.; Hagfeldt, A.; Graetzel, M. Enhanced electronic properties in mesoporous TiO_2 via lithium doping for high-efficiency perovskite solar cells. *Nat. Commun.* **2016**, *7*, 10379. [CrossRef] [PubMed]

32. Abdi-Jalebi, M.; Dar, M.I.; Sadhanala, A.; Senanayak, S.P.; Giordano, F.; Zakeeruddin, S.M.; Grätzel, M.; Friend, R.H. Impact of a Mesoporous Titania-Perovskite Interface on the Performance of Hybrid Organic-Inorganic Perovskite Solar Cells. *J. Phys. Chem. Lett.* **2016**, *7*, 3264–3269. [CrossRef] [PubMed]

33. Kaake, L.G.; Barbara, P.F.; Zhu, X.Y. Intrinsic charge trapping in organic and polymeric semiconductors: A physical chemistry perspective. *J. Phys. Chem. Lett.* **2010**, *1*, 628–635. [CrossRef]

34. Guillén, E.; Ramos, F.J.; Anta, J.A.; Ahmad, S. Elucidating transport-recombination mechanisms in perovskite solar cells by small-perturbation techniques. *J. Phys. Chem. C* **2014**, *118*, 22913–22922. [CrossRef]

35. Salado, M.; Oliva-Ramirez, M.; Kazim, S.; González-Elipe, A.R.; Ahmad, S. 1-dimensional TiO_2 nano-forests as photoanodes for efficient and stable perovskite solar cells fabrication. *Nano Energy* **2017**, *35*, 215–222. [CrossRef]

36. Lopez-Santos, C.; Alvarez, R.; Garcia-Valenzuela, A.; Rico, V.; Loeffler, M.; Gonzalez-Elipe, A.R.; Palmero, A. Nanocolumnar association and domain formation in porous thin films grown by evaporation at oblique angles. *Nanotechnology* **2016**, *27*, 395702. [CrossRef] [PubMed]

37. Contreras-Bernal, L.; Salado, M.; Todinova, A.; Calio, L.; Ahmad, S.; Idígoras, J.; Anta, J.A. Origin and Whereabouts of Recombination in Perovskite Solar Cells. *J. Phys. Chem. C* **2017**, *121*, 9705–9713. [CrossRef]

38. Contreras, L.; Idígoras, J.; Todinova, A.; Salado, M.; Kazim, S.; Ahmad, S.; Anta, J.A. Specific cation interactions as the cause of slow dynamics and hysteresis in dye and perovskite solar cells: A small-perturbation study. *Phys. Chem. Chem. Phys.* **2016**, *18*, 31033–31042. [CrossRef] [PubMed]

39. Leong, W.L.; Ooi, Z.-E.; Sabba, D.; Yi, C.; Zakeeruddin, S.M.; Graetzel, M.; Gordon, J.M.; Katz, E.A.; Mathews, N. Identifying Fundamental Limitations in Halide Perovskite Solar Cells. *Adv. Mater.* **2016**, *28*, 2439–2445. [CrossRef] [PubMed]

© 2018 by the authors. Licensee MDPI, Basel, Switzerland. This article is an open access article distributed under the terms and conditions of the Creative Commons Attribution (CC BY) license (http://creativecommons.org/licenses/by/4.0/).

 *materials*

Article

First-Principles Investigation on the Electronic and Mechanical Properties of Cs-Doped CH$_3$NH$_3$PbI$_3$

Dongyan Liu, Shanshan Li, Fang Bian and Xiangying Meng *

College of Sciences, Northeastern University, Shenyang 110819, China; dongyanliuNEU@163.com (D.L.); shanshanLiNEU@163.com (S.L.); fangbianNEU@163.com (F.B.)
* Correspondence: x_y_meng@mail.neu.edu.cn

Received: 2 June 2018; Accepted: 25 June 2018; Published: 5 July 2018

Abstract: Methylammonium lead iodide, CH$_3$NH$_3$PbI$_3$, is currently a front-runner as light absorber in hybrid solar cells. Despite the high conversion efficiency, the stability of CH$_3$NH$_3$PbI$_3$ is still a major obstacle for commercialization application. In this work, the geometry, electronic structure, thermodynamic, and mechanical property of pure and Cs-doped CH$_3$NH$_3$PbI$_3$ have been systematically studied by first-principles calculations within the framework of the density functional theory (DFT). Our studies suggest that the (CH$_3$NH$_3$)$^+$ organic group takes a random orientation in perovskite lattice due to the minor difference of orientation energy. However, the local ordered arrangement of CH$_3$NH$_3$$^+$ is energetic favorable, which causes the formation of electronic dipole domain. The band edge states of pure and Cs-doped CH$_3$NH$_3$PbI$_3$ are determined by (PbI$_6$)$^-$ group, while A-site (CH$_3$NH$_3$)$^+$ or Cs$^+$ influences the structural stability and electronic level through Jahn–Teller effect. It has been demonstrated that a suitable concentration of Cs can enhance both thermodynamic and mechanical stability of CH$_3$NH$_3$PbI$_3$ without deteriorating the conversion efficiency. Accordingly, this work clarifies the nature of electronic and mechanical properties of Cs-doped CH$_3$NH$_3$PbI$_3$, and is conducive to the future design of high efficiency and stable hybrid perovskite photovoltaic materials.

Keywords: perovskite solar cell; DFT calculations; mechanical property; CH$_3$NH$_3$PbI$_3$

---

1. Introduction

As one of the biggest scientific breakthroughs [1], perovskite solar cell technology has been extensively investigated over the last few years. The conversion efficiency of organic–inorganic lead halide perovskite solar cells has been impressively improved from 3.8% in 2009 to 22.7% recently [2–5]. In particular, methylammonium lead halide perovskites (ABX$_3$; A = (CH$_3$NH$_3$)$^+$; B = Pb$^+$; X = Cl, Br, I) are regarded as particularly promising light absorbers because of their outstanding photovoltaic properties. With its continuous increase in conversion efficiency, the stability of CH$_3$NH$_3$PbI$_3$ based solar cells has attracted tremendous attentions for the scale-up of industrial applications [6–8]. Enhancing the operational stability of CH$_3$NH$_3$PbI$_3$ without weakening the conversion efficiency is still a major challenge for perovskite-type solar cells [9–11].

People attempted to control the composition and proportion of dopants by an alloying method to improve the performance of CH$_3$NH$_3$PbI$_3$-based perovskites. For halogen doping, it has been found that mixing Cl/Br in CH$_3$NH$_3$PbI$_3$ can not only realize the continuous tuning of solar absorption, but also improve the carrier mobility and reduce carrier recombination rates [12,13]. However, due to the thermodynamic instability of the solutions, halogen impurities may cause the segregation of CH$_3$NH$_3$PbI$_3$ into iodide-rich minority and bromide-enriched majority domains under light exposure [14].

Besides the substitution of halogens, B-site doping has also been examined. J. Navas et al. performed experimental and theoretical studies on alloying in Pb^{2+} sites with Sn^{2+}, Sr^{2+}, Cd^{2+}, and Ca^{2+}. They pointed out that the Sn^{2+}, Sr^{2+}, and Cd^{2+} did not modify the phase structure [15]. Y. Ogom et al. proved that the Sn/Pb halide-based perovskite solar cells can harvest the light in the area up to 1060 nm. Nevertheless, the instability caused by the B-site impurities would lead to the reduction of open circuit voltage for $CH_3NH_3PbI_3$ solar cell materials [16].

Recently, the substitution for A-site organic group in $CH_3NH_3PbI_3$ has drawn lots of attentions [17–20]. It has been reported that $[CH(NH_2)_2]_x(CH_3NH_3)_{1-x}PbI_3$ has favorable performances in terms of structural stability, and the band gap of the A-site solutions can be tuned between 1.48 eV and 1.57 eV [21–23]. Yi et al. demonstrated that a mixed A-site cation $Cs_x[CH(NH_2)_2]_{1-x}PbI_3$ exemplifies the potential of high efficiency solar cell material [24]. $[CH(NH_2)_2]_{0.85}Cs_{0.15}PbI_3$ solution has shown a better performance and device stability than the plain $CH(NH_2)_2PbI_3$. Compared with the organic $(CH_3NH_3)^+$ cation, the inorganic Cs^+ is far less volatile [25,26]. At present, it seems that A-site doping is an effective scheme to improve stability without degrading the light conversion efficiency of $CH_3NH_3PbI_3$.

In this work, we first carefully studied the of orientation influence of $(CH_3NH_3)^+$ group on the total energy and band gap in $CH_3NH_3PbI_3$. After the determination of ground state geometry of pure and Cs-doped $CH_3NH_3PbI_3$, a comprehensive investigation on the electronic and mechanical properties was performed by first principles calculations. The present study is conducive to the design of perovskite solar cell material with high stability and efficiency.

2. Calculation Method and Model

The general chemical formula for perovskites is ABX_3. In the cubic unit cell of a perovskite, the A-atom sits at cube corner positions (0, 0, 0) in 12-fold coordination, while the B-atom sits at body center position (1/2, 1/2, 1/2) and is surrounded by six X-atoms to form an octahedron group. A-site traditionally is occupied by the inorganic cations, while in the present hybrid perovskite A-site is inhabited by the organic $(CH_3NH_3)^+$ group. The geometry optimization and the electronic structure calculations of $(CH_3NH_3)^+$ group are investigated by Gaussian 98 code [27], and B3LYP [28,29] method in connection with the 6-311++G basis set [30,31] is chosen as the exchange correlation potential.

All the bulk geometry optimization, electronic structure, and mechanical property of pure and Cs-doped $CH_3NH_3PbI_3$ are calculated with ab initio total energy and molecular dynamics program VASP (VASP 5.4.1, Faculty of Physics, University of Vienna, Austria) [32]. Perdew–Burke–Ernzerhof (PBE) pseudopotential with vdw and spin–orbit coupling (SOC) correction is adopted during the calculation [33]. The plane wave kinetic energy cutoff is set to 550 eV and Brillouin-zone integration is performed with a $12 \times 12 \times 12$ Monkhorst–Pack k-point mesh. The convergence tolerance for the total energy and Hellmann–Feynman force during the structural relaxation is set to 10^{-6} eV and 0.01 eV/Å, respectively.

3. Results and Discussion

3.1. Intrinsic Properties of $CH_3NH_3PbI_3$: Structure and Band

Different locations of $(CH_3NH_3)^+$ group in the perovskite lattice would introduce variations in energy and electronic structure due to the intrinsic degree of freedom of the group. Therefore, it is appropriate to start with the study of the nature of $(CH_3NH_3)^+$. The schematic representation of $(CH_3NH_3)^+$ group is shown in Figure 1a. N and C atoms are all tetrahedral coordinated, and the length of C-N bond in $(CH_3NH_3)^+$ group is 1.52 Å, the Van Der Waals volume is 81.67 Å^3, the density is 0.65 g/cm^3, and the effective radius of the group is 2.69 Å. The electronic property calculations show that $(CH_3NH_3)^+$ holds a 2.54 Debye intrinsic dipole moment with the direction from N atom pointing to C atom. As can be seen in the following, the direction of spontaneous polarization of $(CH_3NH_3)^+$ will influence the stability and bandgap of $CH_3NH_3PbI_3$.

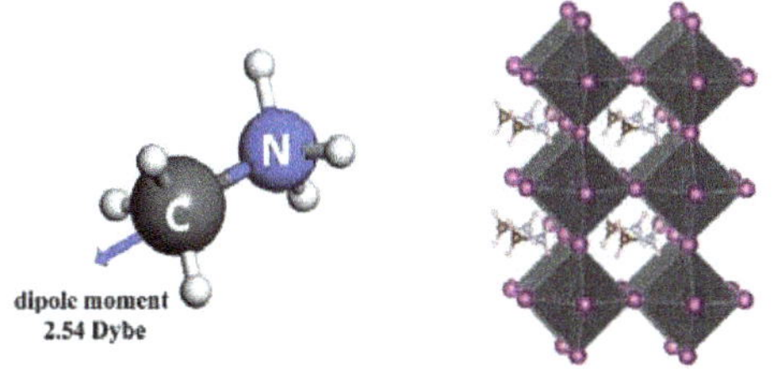

Figure 1. Structure model of (**a**) $(CH_3NH_3)^+$ group (the arrow represents the direction of electric dipole moment in the group), and (**b**) $CH_3NH_3PbI_3$ ground state (the PbI_6 octahedral is rendered).

After the full relaxation of $(CH_3NH_3)^+$ group, we construct $(CH_3NH_3)PbI_3$ unit cell based on the cubic $APbI_3$ inorganic perovskite frame, in which A-site atom is replaced by the $(CH_3NH_3)^+$ group. Different from inorganic cation, the organic $(CH_3NH_3)^+$ group performs a nonspherical shape and can be displayed along different directions in the perovskite lattice [34,35]. In order to determine the energetic favorable configuration, the energy–orientation relationship is studied by fixing the midpoint but rotating the direction of the C-N bond at the center of the cubic cell. Based on the symmetry analysis, there are two independent rotating modes, namely rotating along [110] and [100] directions, for $(CH_3NH_3)^+$ group and the rotation period of each mode is $\pi/2$. The rotating angle is set to zero as the C-N bond lies in the (*ab*) plane.

As shown in Figure 2a, the lowest energies configuration (at 0 K) appears at the position of $\pm 20°$ when rotating along [100] direction. However, the maximum rotation barrier caused by the orientation is about 40 meV, which is close to the thermal energy perturbation at room temperature (26 meV). As a result, $(CH_3NH_3)^+$ group will hold a random orientation in the perovskite lattice at RT. Thus, the calculation results explain the disorder arrangement of the $(CH_3NH_3)^+$ group observed in experiments [36,37].

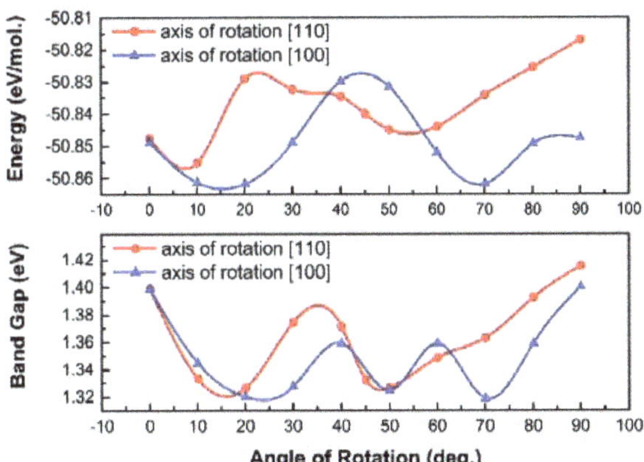

Figure 2. The fluctuation of total energy and bandgap with the orientation of $(CH_3NH_3)^+$ group in the perovskite lattice.

Although the tiny difference in total energy supports the random orientation, we found that the local order location of $(CH_3NH_3)^+$ group is energetic favorable in configuration. Since G.R. Berdiyorov et al. reported that 48 atoms contained supercell is large enough to negligible the finite size effects [38]. We built supercells containing four unit cells and the orientation of $(CH_3NH_3)^+$

group in the adjacent cells is arranged as +−+−, ++−−, and ++++. It is found that the all-aligned ++++ mode is the most stable configuration and ~110 meV lower in total energy than the cross +−+− mode. The local ordered location of $(CH_3NH_3)^+$ is conducive to the establishment of electric dipole domain in the perovskite lattice, which facilitates the separation of photo-generated carriers and leads to the promotion of photoelectric conversion efficiency. This interesting phenomenon has also been investigated by molecular dynamics simulation and experiment [39,40], and the size of the domains is found to be about 100 nm [39,40].

Furthermore, the variation of bandgap with the group orientation is evaluated (As seen in Figure 2b). According to the results, the fluctuation of bandgap caused by the asymmetry of electronic distribution and the electric dipole moment in $(CH_3NH_3)^+$ group is 0.1 eV. It is interesting that the lowest total energy and the minimum bandgap appear at the same point, namely at the position of $\pm 20°$ when rotating along [100] direction. Since that the bandgap shrinks with the intensity of electronic hybridization, it is reasonable to deduce that the system has the lowest energy when the electronic bonding is the strongest.

The point with the lowest energy on the energy–orientation curve corresponds to the ground state of $CH_3NH_3PbI_3$ (Figure 1b). After geometry relaxations, the band structure of ground state in $CH_3NH_3PbI_3$, drawn between high symmetry points of the Brillouin zone, has been illustrated in Figure 3. $CH_3NH_3PbI_3$ is found to have a direct bandgap of 1.68 eV at Γ. This result is well consistent with the reported experimental values [41–43]. The partial electronic densities of state (PDOS) indicates that the upper valence bands (VB) mainly consist of I-$5p$ orbital with weak hybridization to Pb-$5s$ state, and the lower conduction bands (CB) are dominated by Pb-$6p$ orbital. While the $(CH_3NH_3)^+$ group has little contribution to the band edge states. The character of band structure in this work is consistent with recent investigations [44–46]. As a result, the band gap of $CH_3NH_3PbI_3$ will be strongly related to the structure of (PbI_6) inorganic framework. We can thus infer that the influence of $(CH_3NH_3)^+$ group on the electronic structure belongs to Jahn–Teller effect, that is, the change of electronic structure of the system originates from the distortion of (PbI_6) octahedral caused by the size and orientation of $(CH_3NH_3)^+$ group, rather than from the direct participation in the electronic structure of $(CH_3NH_3)^+$. According to our "anion group model" [47], this feature of the electronic structure helps to improve the stability through the element substitution without losing light response and conversion capability.

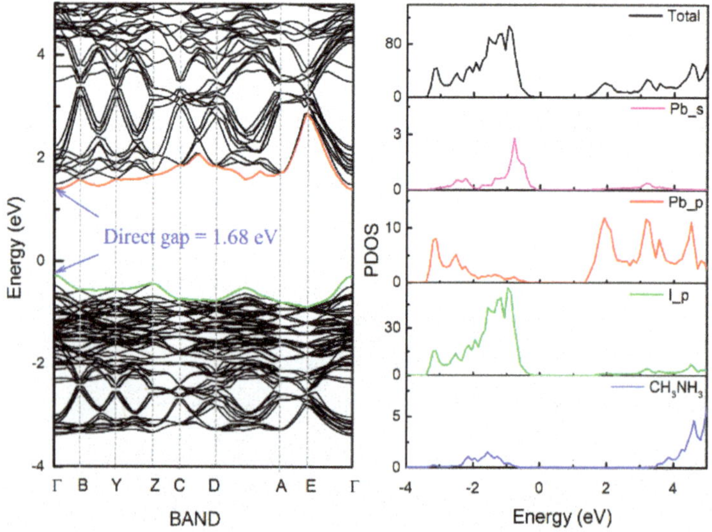

Figure 3. Band level and PDOS of $CH_3NH_3PbI_3$.

3.2. Cs-Doped CH$_3$NH$_3$PbI$_3$: Stability and Electronic Properties

The structure stability of ABX$_3$ perovskites can be estimated by Goldschmidt rule [48], and the tolerance factor t of ABX$_3$ is determined by the expression

$$t = \frac{R_A + R_X}{\sqrt{2}(R_B + R_X)} \tag{1}$$

where R_A, R_B, R_X are the respective effective ionic radii of A, B, and X ions. In general, the perovskite structure is considered to be highly stable when the t-factor is between the range of 0.90~1.00 [49]. Although the above rule has been developed for the oxide perovskites, but the criterion is still valid for the structural stability analysis of inorganic–organic hybrid halide perovskite materials [50,51].

To determine the range of doping concentration of Cs, the Goldschmidt's tolerance factors of Cs$_x$(CH$_3$NH$_3$)$_{1-x}$PbI$_3$ (x from 0 to 1 at an interval of 0.125) are calculated. In our cases, the effective radii of (CH$_3$NH$_3$)$^+$, Pb^{2+}, Cs$^+$, and I$^-$ in the perovskite-type lattice is 2.69 Å, 1.19 Å, 1.88 Å, and 2.20 Å, respectively. As a result, the t-factor is 1.02 for CH$_3$NH$_3$PbI$_3$ and 0.85 for CsPbI$_3$. According the criterion of Goldschmidt, the stability of CH$_3$NH$_3$PbI$_3$ and CsPbI$_3$ is unsatisfactory. For the Cs doped CH$_3$NH$_3$PbI$_3$, we define the average effective ionic radius of A-site cations as $R_{A\text{-}eff} = x \times R_{Cs} + (1 - x) R_{CNH}$. When the concentration x increases from 0 to 1 at an interval of 0.125, the corresponding t-factor at each x is 1.0, 0.99, 0.96, 0.92, 0.91, 0.89, and 0.87 for the Cs-doped CH$_3$NH$_3$PbI$_3$. Thus, from the view of Goldschmidt rule, the Cs-concentration of 12.5 at.% to 62.5 at.% is more desirable. In experiments, it has been found that Cs-doped CH$_3$NH$_3$PbI$_3$ can be successfully synthesized with a high solid solubility [52].

To evaluate the function of Cs-dopant on the structural stability, the 12.5 at.% Cs-doped CH$_3$NH$_3$PbI$_3$ has been modeled by a $2 \times 2 \times 2$ CH$_3$NH$_3$PbI$_3$ supercell with one (CH$_3$NH$_3$)$^+$ group replaced by a Cs atom. In this model we do not consider the disordered orientation of (CH$_3$NH$_3$)$^+$ group. The formation energy of CH$_3$NH$_3$PbI$_3$ and Cs$_{0.125}$(CH$_3$NH$_3$)$_{0.875}$PbI$_3$ are calculated according to the formula

$$E_{formation} = \left(E_{total} - \sum_j n_j E_{ion}^j\right) / N_{total} \tag{3}$$

where E_{total} is the total energy of Cs-doped CH$_3$NH$_3$PbI$_3$, E_{ion}^j is the energy of constituent elements in their respective elemental state, n_j is the number of various constituent elements, and N_{total} is the total number of atoms in the supercell. The results show that the formation energy of Cs$_{0.125}$(CH$_3$NH$_3$)$_{0.875}$PbI$_3$ is 0.02 eV/atom lower than that of the pure CH$_3$NH$_3$PbI$_3$, indicating that incorporating Cs into CH$_3$NH$_3$PbI$_3$ lattice is exothermic and helps to stabilize CH$_3$NH$_3$PbI$_3$. More interestingly, it is found that the original regular (PbI$_6$) octahedral chain has been distorted in the Cs-doped CH$_3$NH$_3$PbI$_3$, as shown in Figure 4. Thus the symmetry breaking of perfect octahedral chain would also lead to the energy reduction.

As the incorporation of Cs will induce a 2.75 Debye net electric dipole moment, we introduce an electric dipole correction term in the electronic structure calculation of Cs$_{0.125}$(CH$_3$NH$_3$)$_{0.875}$PbI$_3$. In Figure 5, energy band diagram clearly shows that the bandgap of 12.5 at.% Cs-doped CH$_3$NH$_3$PbI$_3$ is 1.73 eV, being ~3% wider than that of the pure CH$_3$NH$_3$PbI$_3$. It can be seen that the Cs dopant does not introduce new states at the band edges. This result is consistent with the "anoin group model" in that A-site alkali metal or alkaline earth metal cation in perovskites can hardly influence the electronic state near the Fermi level [47], and this is why the alkali metal doping does not cause the deterioration of photovoltaic performance of CH$_3$NH$_3$PbI$_3$. As a result, A-site participation of alkali metal or alkaline earth metal is a promising way to stabilize the hybrid perovskite photovoltaic materials with reliable photovoltaic properties.

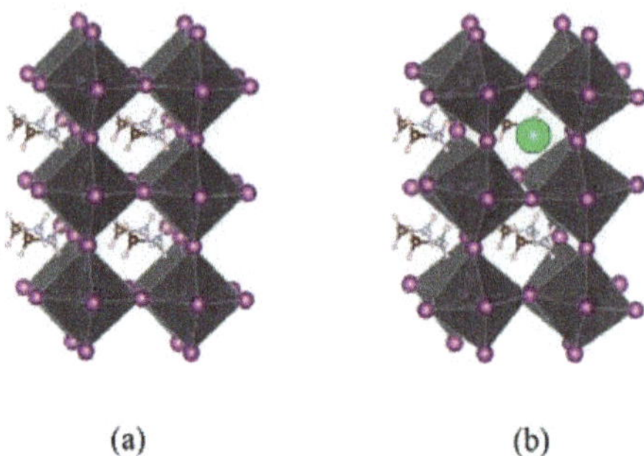

(a)   (b)

Figure 4. Regular (PbI_6) octahedral chain in $CH_3NH_3PbI_3$ (**a**) and distorted chain in 12.5 at.% Cs-doped $CH_3NH_3PbI_3$ (**b**).

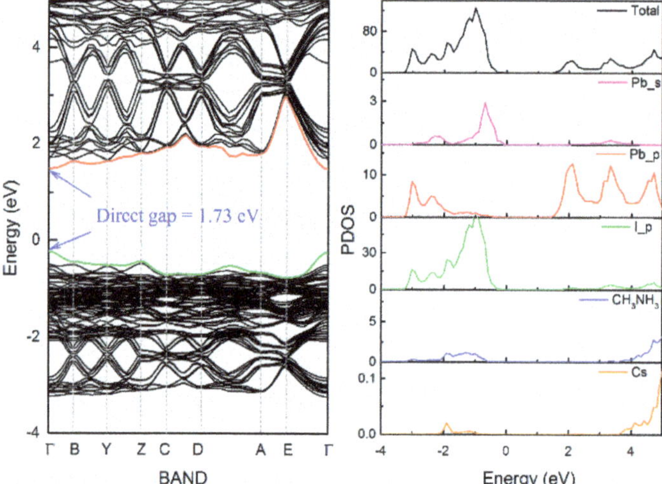

Figure 5. Band level and PDOS of $Cs_{0.125}(CH_3NH_3)_{0.875}PbI_3$.

3.3. Cs-Doped CH₃NH₃PbI₃: Mechanical Properties

It is believed that the mechanical stability of a material is strongly related to its equilibrium elastic modulus and strength performance [53]. The bulk modulus is a measure of resistance to the volume change due to applied pressure, while the shear strength provides information about the resistance of a material against plastic deformation. Thus, the stability of Cs-doped $CH_3NH_3PbI_3$ should be basically understood by these parameters. To obtain elastic moduli, elastic constants were first calculated from the stress–strain relation, then the Voigt–Reuss–Hill (VRH) approximation was applied to the $CH_3NH_3PbI_3$ system, and the effective elastic moduli could be approximated by the arithmetic mean of the Voigt and Reuss limits [54].

In Figure 6, our calculations show that the bulk modulus (B) of pure $CH_3NH_3PbI_3$ is 11.0 GPa, and The shear modulus (G) is 4.9 GPa, which is comparable to the reported experimental value [55,56]. When the Cs-concentration is below 37.5 at.%, both bulk modulus and shear modulus keep going up and the top value of them is higher than that of pure $CH_3NH_3PbI_3$ by 3.5 GPa and 0.6 GPa, respectively. When the Cs-concentration increases further, the corresponding elastic modulus decreases with fluctuation. Our calculations suggest that the Cs-concentration should be controlled in the range of 20~35 at.% if we want to achieve the optimal equilibrium mechanical performance. Due to the limitation of solid solubility, the actual Cs-concentration may be lower than this range. However, the elastic moduli still are higher than that of the pure substance, as shown in Figure 6.

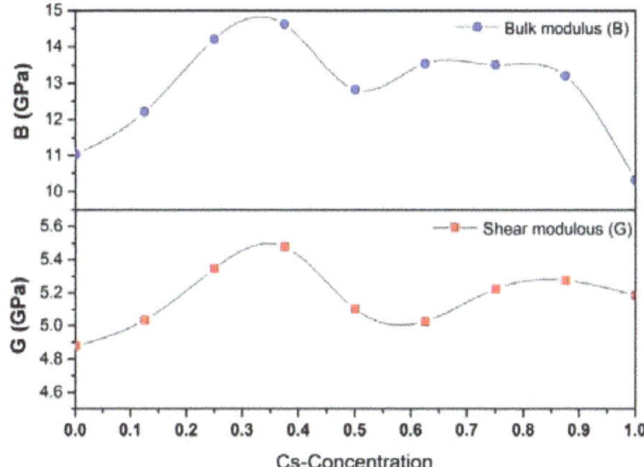

Figure 6. The variation of bulk modulus and shear modulus with Cs-concentration in $CH_3NH_3PbI_3$.

According to the work of Pugh, the material is deemed to be ductile if its B/G ratio is greater than the critical value of 1.75 [57]. The B/G ratio of $CH_3NH_3PbI_3$ is 2.24, while Cs-doped $(CH_3NH_3)PbI_3$ have higher B/G values indicating that those solutions have a better performance in ductility.

The high value of elastic moduli near the equilibrium does not guarantee the high strength. The fracture feature of Cs-doped $(CH_3NH_3)PbI_3$ should be basically understood from the ideal shear strength obtained far from the equilibrium. The ideal shear stress–strain curves of $(CH_3NH_3)PbI_3$ solutions in the (111)<11-2> typical slip system are shown in Figure 7.

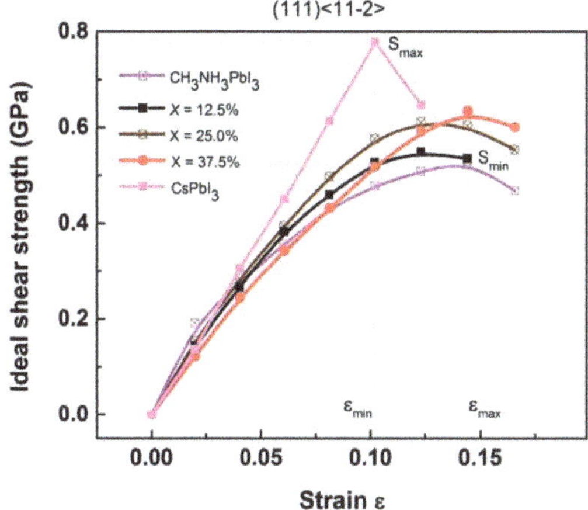

Figure 7. DFT calculated ideal shear stress–strain curves for the $(CH_3NH_3)PbI_3$ solutions.

The strain of deformation-to-failure ε indicates the maximum deformation that the material can bear and thus is the measure of the brittleness. It is found that $CsPbI_3$ has the highest shear strength (0.78 GPa) but the worst strain capacity, showing a feature of high strength and high brittleness. On the country, $(CH_3NH_3)PbI_3$ possess the lowest shear strength (0.52 GPa) but the best strain capacity, displaying a feature of low strength and low brittleness. Below 37.5% Cs-concentration, the fracture performance of solutions lies between the boundaries determined by the pure substances $CsPbI_3$ and $(CH_3NH_3)PbI_3$, that is, the ideal shear strength increases but the maximum bearing deformation decreases. It should be noted that the current strength calculations are strictly related to the defect free lattice, and the obtained ideal might overestimate real shear strengths. Along with the analysis on the equilibrium elastic moduli, we can conclude that below 37.5% Cs-concentration $(CH_3NH_3)PbI_3$ will possess a desirable performance in mechanical properties including stability, hardness, strength, and ductility.

4. Conclusions

Aiming at developing new efficient and stable perovskite solar cell materials, we report a comprehensive theoretical investigation on the geometry, electronic structure, and mechanical property of pure and A-site Cs-doped $CH_3NH_3PbI_3$. The main conclusions are drawn as follows:

(1) The difference in orientation energy of $(CH_3NH_3)^+$ is comparable to the thermal power at room temperature, which causes a random orientation of $(CH_3NH_3)^+$ group in the perovskite lattice.

(2) The local ordered arrangement of $(CH_3NH_3)^+$ is energetic favorable that facilitates the formation of the electronic dipole domain, which helps to improve the separation and lifetime of photo-generated carriers.

(3) The band edge states are dominated by (PbI_6) anion group in $CH_3NH_3PbI_3$. A-site $(CH_3NH_3)^+$ or Cs^+ does not directly participate in the construction of the band edge states, but indirectly influences the structural stability and electronic level through Jahn–Teller effect.

(4) It has been demonstrated that the suitable concentration of Cs can enhance both thermodynamic and mechanical stability of $CH_3NH_3PbI_3$ without deteriorating the conversion efficiency.

(5) Goldschmidt's tolerance factor suggests that the Cs-concentration should be less than 62.5 at.%, while mechanical performance indicates that the optimal Cs-concentration should be less than 37.5%. Below this mark, the mechanical properties including stability, hardness, strength, and ductility will continuously rise with the Cs-concentration.

The adopted research methods, mechanism cognitions and obtained conclusion in the work might help contributing to the future development of efficient and stable hybrid perovskite solar cell materials.

Author Contributions: X.M. and D.L. built the model. D.L., S.L. and B.F. calculated properties of the compounds. All authors contributed to data analysis. X.M. and D.L. wrote the manuscript with input from all authors. X.M. directed the project.

Funding: This work is financially supported by the National Key Research and Development Program of China 2016YFB0701100, the Fundamental Research Funds for the Central Universities (N150502002, N160208001, and N160206002), and National Natural Science Foundation of China (no. 51525101).

Acknowledgments: Dongyan Liu sincerely thanks Haijun Pan at Northeastern University at Qinhuangdao for helpful discussion.

Conflicts of Interest: The authors declare no conflict of interest.

References

1. Science News. Newcomer Juices up the race to harness sunlight. *Science* **2013**, *342*, 1438–1439.

2. Kojima, A.; Teshima, K.; Shirai, Y.; Miyasaka, T. Organometal halide perovskites as visible-light sensitizers for photovoltaic cells. *J. Am. Chem. Soc.* **2009**, *131*, 6050–6051. [CrossRef] [PubMed]

3. Lee, M.M.; Teuscher, J.; Miyasaka, T.; Murakami, T.N.; Snaith, H.J. Efficient hybrid solar cells based on meso-superstructured organometal halide perovskites. *Science* **2012**, *338*, 643–647. [CrossRef] [PubMed]

4. Liu, M.Z.; Johnston, M.B.; Snaith, H.J. Efficient planar heterojunction perovskite solar cells by vapour deposition. *Nature* **2013**, *501*, 395–398. [CrossRef] [PubMed]

5. Best Research-Cell Efficiencies (National Renewable Energy Laboratory, 2018). Available online: https://www.nrel.gov/pv/assets/images/efficiency-chart.png (accessed on 26 June 2018).

6. Conings, B.; Drijkoningen, J.; Gauquelin, N.; Babayigit, A.; D'Haen, J.; D'Oliesleager, L.; Ethirajan, A.; Verbeeck, J.; Manca, J.; Mosconi, E. Intrinsic thermal instability of methylammonium lead trihalide perovskite. *Adv. Energy Mater.* **2015**, *5*. [CrossRef]

7. Christians, J.A.; Herrera, P.A.M.; Kamat, P.V. Transformation of the excited state and photovoltaic efficiency of $CH_3NH_3PbI_3$ perovskite upon controlled exposure to humidified air. *J. Am. Chem. Soc.* **2015**, *137*, 1530–1538. [CrossRef] [PubMed]

8. Graetzel, M.; Janssen, R.A.J.; Mitzi, D.B.; Sargent, E.H. Materials interface engineering for solution-processed photovoltaics. *Nature* **2012**, *488*, 304–312. [CrossRef] [PubMed]

9. Pearson, A.J.; Eperon, G.E.; Hopkinson, P.E.; Habisreutinger, S.N.; Wang, J.T.; Snaith, H.J.; Greenham, N.C. Oxygen Degradation in Mesoporous $Al_2O_3/CH_3NH_3PbI_{3-x}Cl_x$ Perovskite Solar Cells: Kinetics and Mechanisms. *Adv. Energy Mater.* **2016**, *6*. [CrossRef]

10. Alsari, M.; Pearson, A.J.; Wang, J.T.W.; Wang, Z.; Montisci, A.; Greenham, N.C.; Snaith, H.J.; Lilliu, S.; Friend, R.H. Degradation Kinetics of Inverted Perovskite Solar Cells. *Sci. Rep.* **2018**, *8*, 5977. [CrossRef] [PubMed]

11. Alsari, M.; Bikondoa, O.; Bishop, J.; Abdi-Jalebi, M.; Ozer, L.Y.; Hampton, M.; Thompson, P.; Hörantner, M.T.; Mahesh, S.; Greenland, C.; et al. In situ simultaneous photovoltaic and structural evolution of perovskite solar cells during film formation. *Energy Environ. Sci.* **2018**, *11*, 383–393. [CrossRef]

12. Mosconi, E.; Quarti, C.; Ivanovska, T.; Ruani, G.; Angelis, F.D. Structural and electronic properties of organo-halide lead perovskites: A combined IR-spectroscopy and ab initio molecular dynamics investigation. *Phys. Chem. Chem. Phys.* **2014**, *16*, 16137–16144. [CrossRef] [PubMed]

13. Noh, J.H.; Im, S.H.; Heo, J.H.; Mandal, T.N.; Seok, S.I. Chemical management for colorful, efficient, and stable inorganic–organic hybrid nanostructured solar cells. *Nano Lett.* **2013**, *13*, 1764–1769. [CrossRef] [PubMed]

14. Hoke, E.T.; Slotcavage, D.J.; Dohner, E.R.; Bowring, A.R.; Karunadas, H.I.; McGehee, M.D. Reversible photo-induced trap formation in mixed-halide hybrid perovskites for photovoltaics. *Chem. Sci.* **2015**, *6*, 613–617. [CrossRef] [PubMed]

15. Navas, J.; Sánchez-Coronilla, A.; Gallardo, J.J.; Hernández, N.C.; Piñero, J.C.; Alcántara, R.; Lorenzo, C.F.; De los Santos, D.M.; Aguilar, T.; Calleja, J.M. New insights into organic–inorganic hybrid perovskite $CH_3NH_3PbI_3$ nanoparticles. An experimental and theoretical study of doping in Pb^{2+} sites with Sn^{2+}, Sr^{2+}, Cd^{2+} and Ca^{2+}. *Nanoscale* **2015**, *7*, 6216–6229. [CrossRef] [PubMed]

16. Ogomi, Y.; Morita, A.; Tsukamoto, S.; Saitho, T.; Fujikawa, N.; Shen, Q.; Toyoda, T.; Yoshino, K.; Pandey, S.; Ma, T.; et al. $CH_3NH_3Sn_xPb_{(1-x)}I_3$ perovskite solar cells covering up to 1060 nm. *J. Phys. Chem. Lett.* **2014**, *5*, 1004–1011. [CrossRef] [PubMed]

17. Eperon, G.E.; Paternò, G.M.; Sutton, R.J.; Zampetti, A.; Haghighirad, A.A.; Cacialli, F.; Snaith, H.J. Inorganic caesium lead iodide perovskite solar cells. *J. Mater. Chem. A* **2015**, *3*, 19688–19695. [CrossRef]

18. Ahmad, M.; Rehman, G.; Ali, L.; Shafiq, M.; Iqbal, R.; Ahmad, R.; Khan, T.; Asadabad, S.J.; Maqbool, M.; Ahmad, I. Structural, electronic and optical properties of $CsPbX_3$ (X = Cl, Br, I) for energy storage and hybrid solar cell applications. *J. Alloy Compd.* **2017**, *705*, 828–839. [CrossRef]

19. Protesescu, L.; Yakunin, S.; Bodnarchuk, M.I.; Krieg, F.; Caputo, R.; Hendon, C.H.; Yang, R.X.; Walsh, A.; Kovalenko, M.V. Nanocrystals of cesium lead halide perovskites ($CsPbX_3$, X = Cl, Br, and I): Novel optoelectronic materials showing bright emission with wide color gamut. *Nano Lett.* **2015**, *15*, 3692–3696. [CrossRef] [PubMed]

20. Trots, D.M.; Myagkota, S.V. High-temperature structural evolution of caesium and rubidium triiodoplumbates. *J. Phys. Chem. Solids* **2008**, *69*, 2520–2526. [CrossRef]

21. Eperon, G.E.; Beck, C.E.; Snaith, H.J. Cation exchange for thin film lead iodide perovskite interconversion. *Mater. Horiz.* **2016**, *3*, 63–71. [CrossRef]

22. Jeon, N.J.; Noh, J.H.; Yang, W.S.; Kim, Y.C.; Ryu, S.; Seo, J.; Seok, S., II. Compositional engineering of perovskite materials for high-performance solar cells. *Nature* **2015**, *517*, 476–480. [CrossRef] [PubMed]

23. Binek, A.; Hanusch, F.C.; Docampo, P.; Bein, T. Stabilization of the trigonal high-temperature phase of formamidinium lead iodide. *J. Phys. Chem. Lett.* **2015**, *6*, 1249–1253. [CrossRef] [PubMed]

24. Yi, C.; Luo, J.; Meloni, S.; Boziki, A.; Astani, N.A.; Gratzel, C.; Zakeeruddin, S.M.; Rothlisberger, U.; Gratzel, M. Entropic stabilization of mixed A-cation ABX_3 metal halide perovskites for high performance perovskite solar cells. *Energy Environ. Sci.* **2016**, *9*, 656–662. [CrossRef]

25. Li, Z.; Yang, M.; Park, J.S.; Wei, S.H.; Berry, J.J.; Zhu, K. Stabilizing perovskite structures by tuning tolerance factor: Form ation of formamidinium and cesium lead iodide solid-state alloys. *Chem. Mater.* **2016**, *28*, 284–292. [CrossRef]

26. Kulbak, M.; Cahen, D.; Hodes, G. How important is the organic part of lead halide perovskite photovoltaic cells? Efficient $CsPbBr_3$ cells. *J. Phys. Chem. Lett.* **2015**, *6*, 2452–2456. [CrossRef] [PubMed]

27. Frisch, M.J.; Trucks, G.W.; Schlegel, H.B.; Scuseria, G.E.; Robb, M.A.; Cheeseman, J.R.; Zakrzewski, V.G.; Montgomery, J.A.; Stratmann, R.E.; Burant, J.C.; et al. *Gaussian 98*; Gaussian Inc.: Pittsburgh, PA, USA, 1998.

28. Becke, A.D. Density functional thermochemistry. III. The role of exact exchange. *J. Chem. Phys.* **1993**, *98*, 5648–5652. [CrossRef]

29. Lee, C.; Yang, W.; Parr, R.G. Development of the Colle-Salvetti correlation-energy formula into a functional of the electron density. *Phys. Rev. B* **1988**, *37*, 785. [CrossRef]

30. McLean, A.D.; Chandler, G.S. Contracted Gaussian basis sets for molecular calculations. I. Second row atoms, $Z = 11$–18. *J. Chem. Phys.* **1980**, *72*, 5639–5648. [CrossRef]

31. Perdew, J.; Burke, K.; Ernzerhof, M. Generalized gradient approximation made simple. *Phys. Rev. Lett.* **1996**, *77*, 3865. [CrossRef] [PubMed]

32. Kresse, G.; Furthmuller, J. Efficient iterative schemes for ab initio total-energy calculations using a plane-wave basis set. *Phys. Rev. B* **1996**, *54*, 11169. [CrossRef]

33. Krishnan, R.; Binkley, J.S.; Seeger, R.; Pople, J.A. Self-consistent molecular orbital methods. XX. A basis set for correlated wave functions. *J. Chem. Phys.* **1980**, *72*, 650–654. [CrossRef]

34. Bakulin, A.A.; Selig, O.; Bakker, H.J.; Rezus, Y.L.; Muller, C.; Glaser, T.; Lovrincic, R.; Sun, Z.; Chen, Z.; Walsh, A.; et al. Real-time observation of organic cation reorientation in methylammonium lead iodide perovskites. *J. Phys. Chem. Lett.* **2015**, *6*, 3663–3669. [CrossRef] [PubMed]

35. Frost, J.M.; Butler, K.T.; Brivio, F.; Hendon, C.H.; Schilfgaarde, M.V.; Walsh, A. Atomistic origins of high-performance in hybrid halide perovskite solar cells. *Nano Lett.* **2014**, *14*, 2584–2590. [CrossRef] [PubMed]

36. La-o-vorakiat, C.; Salim, T.; Kadro, J.; Khuc, M.; Haselsberger, R.; Cheng, L.; Xia, H.; Gurzadyan, G.G.; Su, H.; Lam, Y.M.; et al. Elucidating the role of disorder and free-carrier recombination kinetics in $CH_3NH_3PbI_3$ perovskite films. *Nat. Commun.* **2015**, *6*, 7903. [CrossRef] [PubMed]

37. Stroppa, A.; Quarti, C.; De Angelis, F.; Picozzi, S. Ferroelectric polarization of $CH_3NH_3PbI_3$: A detailed study based on density functional theory and symmetry mode analysis. *J. Phys. Chem. Lett.* **2015**, *6*, 2223–2231. [CrossRef] [PubMed]

38. Berdiyorov, G.R.; Madjet, M.E.; El-Mellouhi, F.; Peeters, F.M. Effect of crystal structure on the electronic transport properties of the organometallic perovskite $CH_3NH_3PbI_3$. *Sol. Energy Mater. Sol. Cells* **2016**, *148*, 60–66. [CrossRef]

39. Frost, J.M.; Walsh, A. What is moving in hybrid halide Perovskite solar cells? *Acc. Chem. Res.* **2016**, *49*, 528–535. [CrossRef] [PubMed]

40. Kutes, Y.; Ye, L.; Zhou, Y.; Pang, S.; Huey, B.D.; Padture, N.P. Direct observation of ferroelectric domains in solution-processed $CH_3NH_3PbI_3$ Perovskite Thin Films. *J. Phys. Chem. Lett.* **2014**, *5*, 3335–3339. [CrossRef] [PubMed]

41. Jeng, J.Y.; Chiang, Y.F.; Lee, M.H.; Peng, S.R.; Guo, T.F.; Chen, P.; Wen, T.C. $CH_3NH_3PbI_3$ perovskite/fullerene planar heterojunction hybrid solar cells. *Adv. Mater.* **2013**, *25*, 3727–3732. [CrossRef] [PubMed]

42. Qiu, J.; Qiu, Y.; Yan, K.; Zhong, M.; Mu, C.; Yan, H.; Yang, S. All-solid-state hybrid solar cells based on a new organometal halideperovskite sensitizer and one-dimensional TiO_2 nanowire arrays. *Nanoscale* **2013**, *5*, 3245–3248. [CrossRef] [PubMed]

43. Umari, P.; Mosconi, E.; Filippo, D. Relativistic GW calculations on $CH_3NH_3PbI_3$ and $CH_3NH_3SnI_3$ perovskites for solar cell applications. *Sci. Rep.* **2014**, *4*, 4467. [CrossRef] [PubMed]

44. Mosconi, E.; Amat, A.; Nazeeruddin, M.; Gratzel, M.; Angelis, F.D. First-principles modeling of mixed halide organometal perovskites for photovoltaic applications. *J. Phys. Chem. C* **2013**, *117*, 13902–13913. [CrossRef]

45. Umebayashi, T.; Asai, K.; Kondo, T.; Nakao, A. Electronic structures of lead iodide based low-dimensional crystals. *Phys. Rev. B* **2003**, *67*, 155405. [CrossRef]

46. Brivio, F.; Walker, A.B.; Walsh, A. Structural and electronic properties of hybrid perovskites for high-efficiency thin-film photovoltaics from first-principles. *APL Mater.* **2013**, *1*, 042111. [CrossRef]
47. Meng, X.Y.; Liu, D.Y.; Qin, G.W. Band engineering of multicomponent semiconductors: A general theoretical model on the anion group. *Energy Environ. Sci.* **2018**, *11*, 692–701. [CrossRef]
48. Goldschmidt, V.M. Crystal structure and chemical correlation. *Ber. Dtsch. Chem. Ges.* **1927**, *60*, 1263–1268. [CrossRef]
49. Goldschmidt, V.M. Die Gesetze der Krystallochemie. *Naturwissenschaften* **1926**, *14*, 477–485. [CrossRef]
50. Kieslich, G.; Sun, S.; Cheetham, A.K. Solid-state principles applied to organic–inorganic perovskites: New tricks for an old dog. *Chem. Sci.* **2014**, *5*, 4712–4715. [CrossRef]
51. Stoumpos, C.C.; Kanatzidis, M.G. The renaissance of halide perovskites and their evolution as emerging semiconductors. *Acc. Chem. Res.* **2015**, *48*, 2791–2802. [CrossRef] [PubMed]
52. Niemann, R.G.; Gouda, L.; Hu, J.; Tirosh, S.; Gottesman, R.; Cameron, P.J.; Zaban, A. Cs^+ incorporation into $CH_3NH_3PbI_3$ perovskite: Substitution limit and stability enhancement. *J. Mater. Chem. A* **2016**, *4*, 17819–17827. [CrossRef]
53. Born, M.; Huang, K. *Dynamical Theory of Crystal Lattices*; Clarendon Press: Oxford, UK, 1954.
54. Hill, R. The elastic behaviour of a crystalline aggregate. *Proc. Phys. Soc. A* **1952**, *65*, 349–354. [CrossRef]
55. Beecher, A.N.; Semonin, O.E.; Skelton, J.M.; Frost, J.M.; Terban, M.W.; Zhai, H.W.; Alatas, A.; Owen, J.S.; Walsh, A.; Billinge, S.J.L. Direct observation of dynamic symmetry breaking above room temperature in methylammonium lead iodide perovskite. *ACS Energy Lett.* **2016**, *1*, 880–887. [CrossRef]
56. Rakita, Y.; Cohen, S.R.; Kedem, N.K.; Hodes, G.; Cahen, D. Mechanical properties of $APbX_3$ (*A* = Cs or CH_3NH_3; *X* = I or Br) perovskite single crystals. *MRS Commun.* **2015**, *5*, 623–629. [CrossRef]
57. Pugh, S.F. Relations between the elastic moduli and the plastic properties of polycrystalline pure metals. *Philos. Mag. Ser.* **1954**, *45*, 823–843. [CrossRef]

© 2018 by the authors. Licensee MDPI, Basel, Switzerland. This article is an open access article distributed under the terms and conditions of the Creative Commons Attribution (CC BY) license (http://creativecommons.org/licenses/by/4.0/).

 *materials*

Article

Photoactive ZnO Materials for Solar Light-Induced Cu$_x$O-ZnO Catalyst Preparation

Magdalena Brzezińska [1,2], Patricia García-Muñoz [2], Agnieszka M. Ruppert [1] and Nicolas Keller [2,*]

[1] Institute of General and Ecological Chemistry, Faculty of Chemistry, Lodz University of Technology, 90-924 Łódź, Poland; brzezinskamm@gmail.com (M.B.); agnieszka.ruppert@p.lodz.pl (A.M.R.)

[2] Institut de Chimie et Procédés pour l'Energie, l'Environnement et la Santé, CNRS/University of Strasbourg, 67087 Strasbourg, France; garciamunoz@unistra.fr

* Correspondence: nkeller@unistra.fr; Tel.: +33-3-6885-2811

Received: 22 September 2018; Accepted: 11 November 2018; Published: 13 November 2018

Abstract: In this work, the solar light-induced redox photoactivity of ZnO semiconductor material was used to prepare Cu$_x$O-ZnO composite catalysts at room temperature with a control of the chemical state of the copper oxide phase. Cu$_2^{(I)}$O-ZnO and Cu$^{(II)}$O-ZnO composite catalysts were prepared by using Cu(acac)$_2$ in tetrahydrofuran-water and Cu(NO$_3$)$_2$ in water as metallic precursor, respectively. Prior to the implementation of the photon-assisted synthesis method, the most efficient photoactive ZnO material was selected from among different ZnO materials prepared by the low temperature polyol and precipitation methods with carbonates and carbamates as precipitation agents. The photocatalytic degradation of the 4-chlorophenol compound in water under simulated solar light was taken as a model reaction. The ZnO support materials were characterized by X-ray diffraction (XRD), surface area and porosimetry measurements, thermogravimetric analysis (TGA), scanning electron microscopy (SEM) and transmission electron microscopy (TEM), and the synthesis method strongly influenced their photoactivity in terms of 4-chlorophenol degradation and of total organic carbon removal. The most photoactive ZnO material was prepared by precipitation with carbonates and calcined at 300 °C, benefitting from a high specific surface area and a small mean crystallite size for achieving a complete 4-chlorophenol mineralization within 70 min of reaction, with minimum Zn^{2+} released to the solution. Besides thermal catalysis applications, this work has opened a new route for the facile synthesis of Cu$_2$O-ZnO heterojunction photocatalysts that could take place under solar light of the heterojunction built between the *p*-type semi-conductor Cu$_2$O with direct visible light band gap and the ZnO semiconductor phase.

Keywords: ZnO; photo-oxidation; 4-chlorophenol; Cu$_x$O-ZnO catalyst; photodeposition

---

1. Introduction

Zinc oxide (ZnO) nanostructures are materials with potential applications in many fields of nanotechnology, due to the large variety of nanometric structures or architectures that can be synthesized [1]. Besides numerous applications in optoelectronics, electronics, laser technology, converters and energy generators or as gas sensors [1–6], zinc oxide (ZnO) nanostructures have received special attention for being used directly as (photo) catalyst or as catalyst support.

Indeed, ZnO is a II–VI compound semi-conductor that has emerged as a promising candidate for being used as heterogeneous photocatalyst under near-UV irradiation in environmental applications, such as the removal of a large range of organic and inorganic contaminants from environmental water and wastewater, including some of the most toxic and refractory molecules in water like pesticides, herbicides, and dyes [7]. It is characterized by a direct wide band gap close to that of anatase TiO$_2$ (3.2–3.37 eV), a suitable location of both conduction and valence bands, strong oxidation ability, low cost, abundance and a large exciton binding energy of 60 meV at room temperature [1,8–11] so that exciton emission processes can persist at or even above room temperature. It has already been reported that the electron lifetime can be significantly higher and that the recombination rate can be lower in ZnO in comparison to TiO_2, making ZnO attractive and worth investigating for photocatalytic applications [12]. Further, the photocatalytic activity of ZnO can be enhanced by designing ZnO-supported metal nanoparticle photocatalysts, with promising results mainly in the case [13] of Au and Ag [14–19]. In such a hybrid metal/ZnO nanostructure, the supported nanoparticles are proposed (i) to act as a sink for photoinduced electrons, so that efficient charge separation could be promoted at the semi-conductor/metal interface with efficient interfacial charge transfer, or (ii) to induce plasmonic effects either through the direct injection of hot (excited) electrons from the metal to the conduction band of ZnO thanks to intimate electrochemical contact between the plasmonic particle and the semiconductor, or through a near-field enhancement mechanism with overlap between the plasmon wavelength and the photocatalyst absorption.

ZnO nanostructures are also widely studied as catalyst support in several reactions of high fundamental or applicative interest, such as methanol oxidation [20]; low-temperature methanol synthesis [21,22]; production of hydrogen from methanol by steam reforming, partial oxidation, or a combination thereof [23]; steam reforming of ethanol; glycerol hydrogenolysis [24]; various transesterification reactions [25]; or, selective hydrogenation reactions [26–30]. The active metals or metal oxides have mainly included Cu, Pd, Au, Ag, Co, Ni, Rh, Ir and Pt, and were usually stabilized on the ZnO support by ion exchange, wet or incipient wetness impregnation, or (co)-precipitation [31].

The high interest of using zinc oxide as a support results from its valuable physico-chemical properties. Hydrogenation reactions can benefit from strong metal-support interactions and from the formation of alloys with the supported metals in reductive atmosphere. This behavior is responsible for the modification of the electronic properties of the supported active metals, which in turn results in significant activity and selectivity enhancements in many catalytic reactions [32]. Activity improvement can be also associated with the existence of a hydrogen spillover phenomena taking place in the case of zinc oxide-supported metal catalysts [33,34].

The most-used methods for preparing supported metal catalysts implement consecutive operations, beginning with the adsorption of the metal precursor on the support. The reduction step consists mostly of a thermal treatment under hydrogen, or in a chemical reduction in a solvent with, for example, hydrazine or sodium borohydrate reactants. The catalysts can suffer from heterogeneous supported nanoparticle size distributions, from unwanted temperature-activated reactions between the support and the metal precursor, and from limitations in terms of metal contents.

It has been demonstrated that the synthesis method is playing a key role in the preparation of ZnO materials with varied bulk and surface physico-chemical properties, and many chemical methods have been reported for synthesizing ZnO nanostructures, including notably mechanico-chemical processes; precipitation; sol-gel, solvothermal and hydrothermal methods; methods using an emulsion or microemulsion environment; sonochemical; or, microwave-based methods [4–6,11].

Therefore, the aim of this article is two-fold. First, it aims at selecting an appropriate preparation method for synthesizing photoactive ZnO under solar light. This has been performed by studying the influence of the synthesis method on ZnO photoactivity, taking the photocatalytic degradation of 4-chlorophenol in water under solar light as a model reaction. Further, the solar light-induced redox photoactivity developed by ZnO has been used for room-temperature preparation of Cu_xO-ZnO catalysts that are of interest in fundamental and applicative reactions, and that do not require the use of any thermal treatment, or of any gaseous or liquid reductant for controlling the Cu chemical state.

2. Materials and Methods

2.1. Synthesis of ZnO Materials

2.1.1. Polyol Method

In the polyol synthesis, 1.5 g of zinc(II) acetate dihydrate (Zn(OAc)$_2$·2H$_2$O, Sigma-Aldrich, Saint Louis, MO, USA, ACS reagent, ≥98%) was introduced into 50 mL of propane-1,3-diol solvent (C$_3$H$_8$O$_2$, Sigma-Aldrich, 98%). The obtained mixture was kept under continuous stirring and under reflux at 160 °C for 15 min or 60 min. Afterwards, the precipitate obtained was cooled, centrifuged for 20 min at 3500 rpm, and finally washed and filtrated under vacuum several times with absolute ethanol (Sigma-Aldrich). The synthesized powder was dried at 100 °C for 12 h. The samples obtained were labelled as ZnO-P-15 and ZnO-P-60 according to the reflux duration.

2.1.2. Precipitation with Na$_2$CO$_3$

In the precipitation synthesis method, 1.75 g of zinc(II) acetate dihydrate and 0.84 g of sodium carbonate (Na$_2$CO$_3$, Sigma-Aldrich, 99.5%) were dissolved under stirring in 50 mL of distilled water, respectively. Both aqueous solutions were mixed, and the obtained precipitate was aged at room temperature in the mother liquor for 24 h under continuous stirring. The suspension was further centrifuged for 30 min at 3500 rpm, and finally washed and filtrated under vacuum with distilled water. The resulting powder was dried at 100 °C for 12 h and subsequently calcined at a temperature of 300 °C to 500 °C for 2 h with a 10 °C/min heating rate, leading to the ZnO-C material series.

2.1.3. Precipitation with NH$_2$CO$_2$NH$_4$

Quantities of 10.98 g of zinc(II) acetate dihydrate and 4.29 g of ammonium carbamate (NH$_2$CO$_2$NH$_4$, Sigma Aldrich, 99%) were dissolved under stirring in 50 mL of distilled water. Both aqueous solutions were mixed, and the obtained precipitate was aged at room temperature in the mother liquor for 30 min under continuous stirring. The suspension was further centrifuged for 30 min at 3500 rpm, and finally washed and filtrated under vacuum with distilled water. The resulting powder was dried at 100 °C for 12 h and subsequently calcined at a temperature of 400 °C to 500 °C for 2 h with a 10 °C/min heating rate, leading, e.g., to the ZnO-c material series.

2.2. Photon-Assisted Preparation of Cu-ZnO Catalysts

The photon-assisted preparation of Cu-ZnO catalysts was performed by irradiating with solar light a suspension of the ZnO support containing Cu(acac)$_2$ or Cu(NO$_3$)$_2$ as metallic precursors. The irradiation was provided by an Atlas Suntest XLS+ reaction chamber equipped with a Xenon arc lamp adjusted to a 500 W/m^2 irradiance (with a 30 W/m^2 UV-A content) (320–800 nm wavelength range, ICH Q1B guidelines). The amount of copper precursor used was adjusted to target a Cu content of 10 wt % on the ZnO support.

2.2.1. Cu(acac)$_2$ in THF-H$_2$O Solvent

The dissolution of 20.8 mg of copper(II) acetylacetonate (Sigma-Aldrich, 97%) was achieved for 30 min at 60 °C under stirring in tetrahydrofuran (THF, Sigma-Aldrich, 99%)-water mixture with a 2:9 (*v*/*v*) ratio, before 45 mg of the ZnO support was dispersed under stirring in 100 mL of copper solution in a beaker-type glass reactor at a 0.21 g/L concentration. The suspension was first stirred in the dark for 0.5 h at 60 °C for establishing the adsorption/desorption equilibrium, before the photon-assisted synthesis was performed under stirring at 60 °C under solar light for 90 min. The synthesis was monitored by UV-vis spectrophotometry by following the disappearance of the absorption peak at λ = 245 nm. The samples were washed and filtrated under vacuum several times with distilled water, and finally dried at 100 °C for 1 h [13].

2.2.2. Cu(NO$_3$)$_2$ in H$_2$O Solvent

The protocol was similar to that followed with Cu(acac)$_2$. In this case, the dissolution of 38 mg of Cu(II) nitrate trihydrate (Sigma-Aldrich, p.a) was performed in water at room temperature, before 90 mg of ZnO was dispersed under stirring in 100 mL of copper solution in a beaker-type glass reactor at a 0.38 g/L concentration. Prior to irradiation, the suspension was stirred in the dark for 30 min at 60 °C to ensure the establishment of the adsorption/desorption equilibrium, before the photon-assisted synthesis was performed under stirring under solar light for 2 h. The synthesis was monitored by UV-vis spectrophotometry by following the disappearance of the absorption peak at λ = 800 nm. The samples were washed and filtrated under vacuum several times with distilled water, and finally dried at 100 °C for 1 h.

2.3. Characterization Techniques

The crystallographic structure of the powders was characterized by X-ray diffraction patterns (XRD) recorded on a D8 Advance Bruker diffractometer in a θ/θ mode and using the K$_{\alpha 1}$ radiation of a Cu anticathode (λ = 1.5406 Å).

The surface area measurements were carried out on a Micrometrics Tristar 3000 using N$_2$ as adsorbent at -196 °C with a prior outgassing at 100 °C overnight in order to desorb the impurities or moisture. The Brunauer-Emmett-Teller (BET) specific surface area was calculated from the N$_2$ adsorption isotherm.

Scanning electron microscopy (SEM) was performed in secondary electron mode on a JSM-6700 F FEG microscope (JEOL Ltd., Tokyo, Japan).

Transmission electron microscopy (TEM) was performed using a JEOL 2100F (JEOL Ltd., Tokyo, Japan) with a point resolution of 0.2 nm. A Mo grid was used for performing Energy-dispersive X-ray spectroscopy (EDS) analysis of Cu-ZnO samples. The interplanar spacings were calculated using ImageJ software (National Institutes of Health, Bethesda, MD, USA).

Thermogravimetric analysis (TGA) was carried out with a 20% (v/v) O$_2$/N$_2$ mixture at a flow rate of 40 mL/min at a heating rate of 10 °C/min from 25 °C to 600 °C using a Q 5000 thermoanalyzer from TA instrument (New Castle, DE, USA).

The copper content in the catalysts was determined by chemical analysis after a microwave-assisted acidic dissolution in aqua regia at 185 °C under autogenic pressure. Inductively coupled plasma optical emission spectroscopy (ICP-OES) was carried out on an Optima 7000 DV spectrometer (Perkin Elmer, Waltham, MA, USA) at the Analysis Platform of IPHC-Strasbourg, France.

2.4. Evaluation of the Photocatalytic Efficiency

The experiments were carried out in an Atlas Suntest XLS+ reaction chamber equipped with a Xenon arc lamp adjusted to a 500 W/m^2 irradiance (with a 30 W/m^2 UV-A content) (320–800 nm wavelength range, ICH Q1B guidelines). The tests were performed with a reactor volume of 500 mL and a 4 Cl-phenol concentration of 25 mg L^{-1}, corresponding to a carbon load of 14.8 ppm. The ZnO catalyst load was 1 g L^{-1}. At each time interval, 20 mL of solution was sampled and then filtered through 0.20 μm porosity filter to remove the photocatalyst powder, if any, before the concentration of 4-chlorophenol was determined by UV-visible spectrophotometry (Cary 100 scan) by monitoring the disappearance of the main absorption peak at λ = 224 nm, and total organic carbon (TOC) was measured with a TOC-L analyzer (Shimadzu, Kyoto, Japan) to determine the organic carbon load.

A procedure was established for photocatalytically cleaning the ZnO materials under UV-A or solar light prior to the chlorophenol removal, consisting in the continuous stirring of the as-synthesized ZnO catalyst in ultrapure Milli-Q water under either simulated solar light for 2 h or UV-A light for 16 h. UV-A light was provided by blacklight lamps (PL-L 24W/10/4P, Philips, Amsterdam, The Netherlands) emitting at 365 nm with an irradiance of 60 W/m^2, while the UV-A fraction of the simulated solar light corresponded to a 30 W/m^2 irradiance, as measured using a wide-band RPS900-W rapid portable spectroradiometer from International Light Technology (ILS, Peabody, MA, USA).

3. Results

3.1. Characterization of ZnO Materials

The main physico-chemical properties of the ZnO materials are reported in Table 1. Figure 1a,b shows the XRD patterns of the dried ZnO precursors and of the final ZnO materials synthesized via the polyol and the precipitation routes.

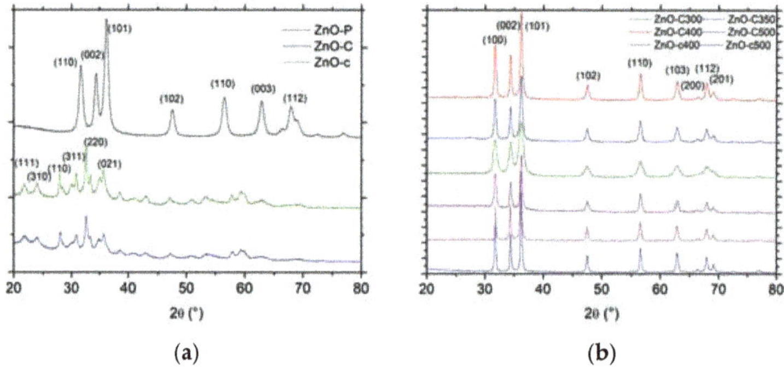

Figure 1. X-ray diffraction patterns (XRD) patterns of (**a**) the dried materials and (**b**) the calcined materials. JCPDS card No. 00-036-1451 and No. 19-1458 for ZnO and the zinc carbonate hydroxide phases, respectively.

In the case of the polyol method, the XRD pattern shown in Figure 1a displayed the diffraction peaks characteristic of ZnO crystallized in the hexagonal wurtzite structure and corresponding to the diffraction of the (100), (002), (101), (102), (110), and (103) planes for the most intense peaks (P63mc space group, JCPDS 00-036-1451) [5,14–16]. As reported for the polyol synthesis, the ZnO material had a good crystallinity without any post-synthesis calcination treatment.

By contrast, the dried material obtained in the case of both precipitation methods consisted of a $Zn_5(CO_3)_2(OH)_6$ zinc carbonate hydroxide, or hydrozincite phase, in the monoclinic structure [17] with the diffraction of the (111), (310), (100), (311), (220) and (021) planes for the most intense peaks (JCPDS 19-1458). Han et al. [18] have proposed that the reactions during the formation of $Zn_5(CO_3)_2(OH)_6$ are likely as follows:

$$Zn(CH_3COO)_2 \rightarrow Zn^{2+} + 2[CH_3COO]^- \tag{1}$$

$$(CH_2)_6N_4 + 10H_2O \rightarrow 6HCHO + 4OH^- + 4[NH_4]^+ \tag{2}$$

$$HCHO + O_2 + 2OH^- \rightarrow CO_3^{2-} + 2H_2O \tag{3}$$

$$5Zn^{2+} + 2CO_3^{2-} + 6OH^- \rightarrow Zn_2(CO_3)_2(OH)_6 \tag{4}$$

The zinc carbonate hydroxide phase being thermally stable below 200 °C, its decomposition into ZnO was evidenced by XRD characterization and TGA analysis in Figures 1 and 2, respectively. After calcination of the zinc carbonate hydroxide at a temperature higher than 300 °C, the XRD pattern exhibited the diffraction peaks corresponding to hexagonal ZnO, while, according to the weight loss of $28 \pm 1\%$ observed in TGA, the decomposition process can be described as follows, corresponding to a theoretical weight loss of 28.4%:

$$Zn_2(CO_3)_2(OH)_6 \rightarrow 5ZnO + 2CO_2 + 3H_2O \tag{5}$$

The absence of any weight loss during the TGA of the calcined ZnO materials confirmed the efficiency of the thermal decomposition process. By contrast, a very small weight loss of 6% was observed with ZnO prepared via the polyol method, suggesting the residual presence of propane 1-3 diol solvent or acetate species [19].

In agreement with the previous works of Fkiri et al. [14], the ZnO materials synthesized via the polyol method displayed a good crystallinity without any post-synthesis heat treatment, and exhibited the smallest average crystallite size of 7 nm and 11 nm, for an aging duration of 15 min and 60 min, respectively. The ZnO materials prepared via both precipitation routes had a larger mean crystallite size, as a result of the necessary calcination treatment at temperatures ranging from 300 °C to 500 °C. The use of carbonates as precipitating agent allowed maintaining a smaller mean crystallite size in the 10–20 nm range, while the mean crystallite size strongly increased to 28–30 nm in the case of the carbamate agent. For a similar calcination temperature, ZnO prepared with carbamates displayed a larger crystallite size than its carbonate counterpart, e.g., 28 nm vs. 20 nm at a temperature of 400 °C.

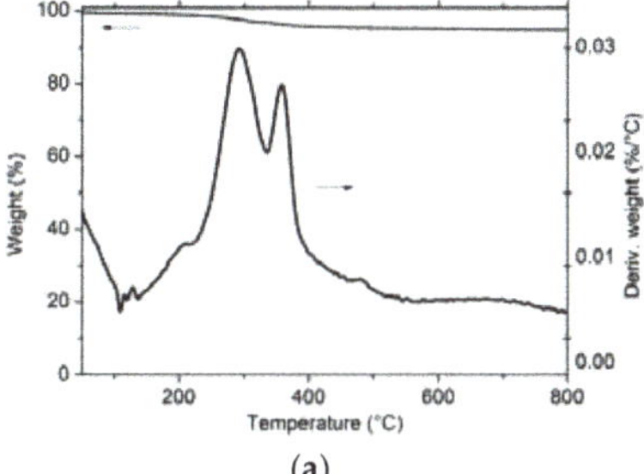
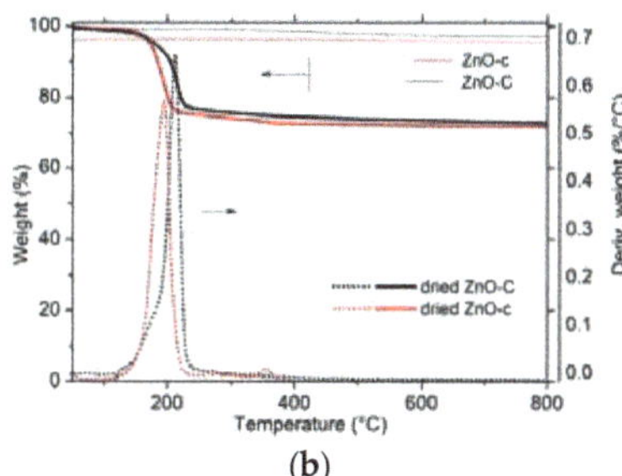

(a)   (b)

Figure 2. Thermogravimetric analysis (TGA) of the materials after the drying step synthesized by (**a**) the polyol method; (**b**) the carbonate-derived precipitation method and the carbamate-derived precipitation method. The weight loss observed with the ZnO-C and ZnO-c materials after calcination at 300 °C and 400 °C, respectively, was reported.

Table 1. Main physico-chemical properties of ZnO materials.

Sample	Mean Crystallite Size (nm) [a]	BET Surface Area (m^2/g)	Pore Volume (cm^3/g)	Average Pore Diameter (nm)
ZnO-P-15	7	78	0.13	5
ZnO-P-60	11	66	0.09	6
ZnO-C-300	10	68	0.45	21
ZnO-C-350	22	41	0.24	21
ZnO-C-400	20	29	0.21	27
ZnO-C-500	18	29	0.24	30
ZnO-c-400	28	12	0.10	30
ZnO-c-500	30	6	0.05	28
Cu$_2$O-ZnO	13	95	0.68	26
CuO-ZnO	20	33	0.24	29

[a] derived from XRD pattern, as the mean size of the coherently-diffracting domains, derived from the Scherrer equation using the classical assumption of spherical crystallites. The full-width at half-maximum of the diffraction peaks of ZnO (102), (110), and (103) planes was used for the estimation.

Figure 3a shows the N$_2$ adsorption–desorption isotherms of the ZnO materials. The ZnO-P materials displayed IV-type isotherms, characteristic of mainly mesoporous solids with a mainly H2-type hysteresis corresponding to interconnected mesopores with non-uniform shape or size [14]. For ZnO obtained by precipitation, the isotherms were found to be preferentially of type II with H3-type hysteresis. This suggested that the porosity mainly resulted from macropores (or large mesopores) and is characteristic of aggregates or agglomerates of nanoparticles forming slit-shaped pores with non-uniform size and/or shape. This was in agreement with the corresponding pore size distributions shown in Figure 3b, which clearly evidenced the influence of the synthesis method on the mean pore size.

The specific surface area of the ZnO materials ranged from 6 m^2/g to 78 m^2/g without any microporous contribution. Among the different materials, the ZnO-P and ZnO-C-300 samples had the highest specific surface areas, within the 66–78 m^2/g range, in agreement with their smallest mean crystallite sizes of 7–11 nm. The increase in the calcination temperature for carbonate-derived ZnO materials from 300 °C to 500 °C led to a progressive decrease in the surface area from 68 m^2/g to 29 m^2/g, as a result of the slight increase in the mean crystallite size. It is worth noting that the carbamate-derived ZnO samples calcined at 400 °C and 500 °C displayed a lower surface area than their carbonate-derived counterparts, at 12 m^2/g and 6 m^2/g respectively, in agreement with a larger mean crystallite size.

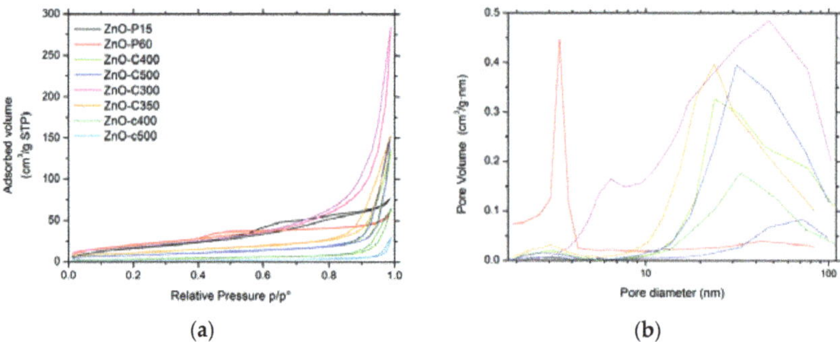

Figure 3. (**a**) N$_2$ adsorption-desorption isotherms and (**b**) pore size distributions of ZnO materials.

The synthesis method had no influence on the band gap of the ZnO materials, graphically estimated at 3.1 eV via the Tauc plot derived from the UV–vis diffuse reflectance analysis, in agreement with the literature (Figure S1b) [14,35]. Therefore, the ZnO materials will be activated by incident photons with wavelengths within the UV-A range, whether the photocatalytic tests are performed under solar light or pure UV-A light.

Figure 4 depicts a selection of scanning electron microscopy (SEM) images of the ZnO materials and evidences that the synthesis method and the nature of the precipitating agent strongly influenced the morphology of the ZnO material. ZnO from polyol synthesis was composed of (100–200 nm) large spherical aggregates of small-size ZnO particles with an average crystallite size around 10 nm, in agreement with that derived from the XRD patterns. The aging duration had no influence on the general morphology of the ZnO-P materials. The interplanar spacing of 0.28 and 0.25 nm shown in Figure 5 was consistent with the (100) and (101) plane of hexagonal (wurtzite) ZnO crystallites.

Transmission electron microscopy (TEM) images of ZnO synthesized via the different methods are shown in Figure 5. Irrespective of the synthesis method, they evidenced interplanar spacings of 0.28 nm and 0.25 nm consistent with the (100) and (101) planes of the hexagonal wurtzite ZnO phase (JCPDS No. 00-036-1451) and confirmed the general morphology of the wurtzite ZnO materials previously observed in SEM images, with crystallite sizes in agreement with those derived from the XRD patterns.

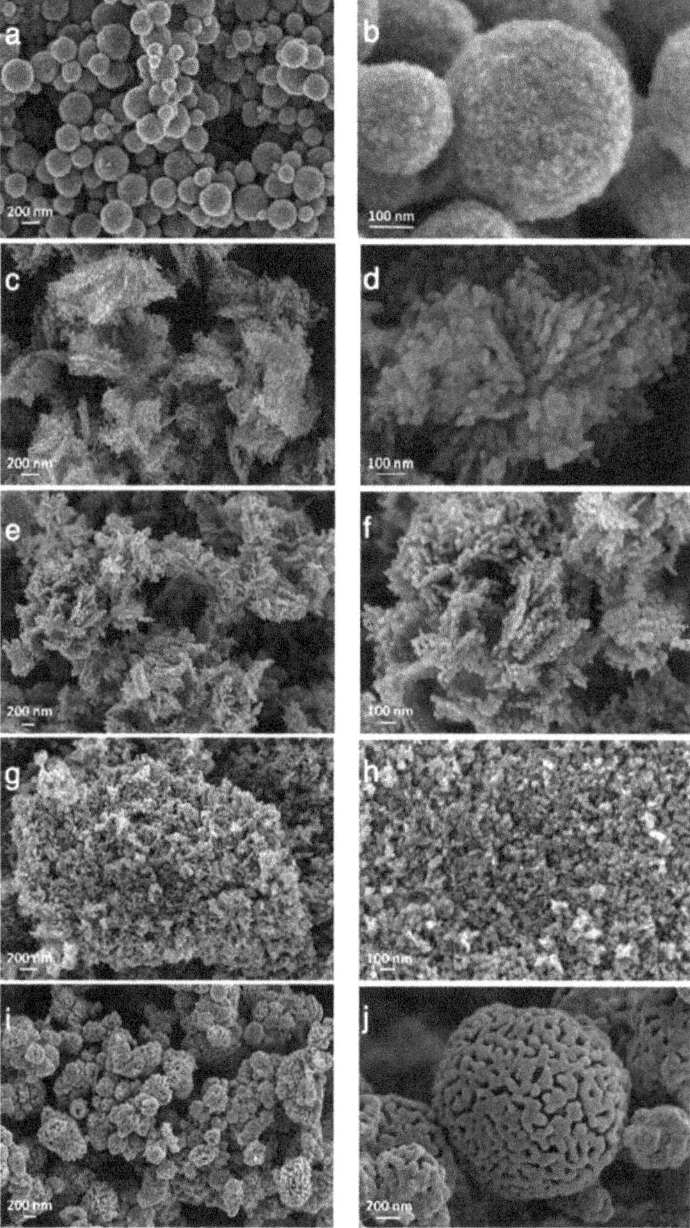

Figure 4. Scanning electron microscopy (SEM) of ZnO materials synthesized through (**a**,**b**) the polyol method (ZnO-P-60); the precipitation with carbonates: (**c**,**d**) ZnO-C300; (**e**,**f**) ZnO-C400; (**g**,**h**) ZnO-C500; and the precipitation with carbamates: (**i**) ZnO-c400 and (**j**) ZnO-c500.

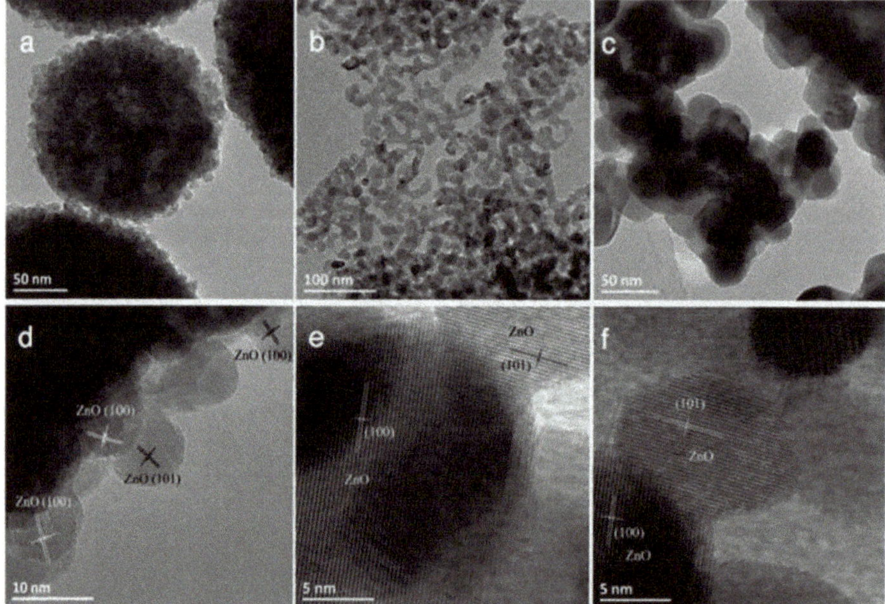

Figure 5. Transmission electron microscopy (TEM) images of ZnO materials synthesized through (**a**,**d**) the polyol method (ZnO-P-60), (**b**,**e**) the precipitation with carbonates (ZnO-C300) and (**c**,**f**) the precipitation with carbamates (ZnO-c400), evidencing the general morphology of the materials and interplanar spacings of 0.28 nm and 0.25 nm consistent with the (100) and (101) planes of hexagonal wurtzite ZnO crystallites.

3.2. Photocatalytic Activity of ZnO

3.2.1. Influence of a Photocatalytic Pre-Cleaning Step

The photocatalytic activity of the as-synthesized ZnO materials was first evaluated in water under solar light using 4-Cl-phenol as test molecule. Two catalysts prepared via the polyol method and the carbonate-through precipitation method were selected for studying the influence of a photocatalytic pre-cleaning step, i.e., the dried ZnO-P-15 (sample not subjected to any calcination treatment) and the calcined ZnO-C-300 sample, respectively (Figure 6).

First, no detectable degradation of chlorophenol was observed for 3 h under solar light, demonstrating that the chlorophenol photolysis can be disregarded, as evidenced also by Eslami et al. and Gaya et al. [36,37].

As depicted in Figure 6a, the pollutant concentration evolution with the time under irradiation showed that both ZnO catalysts were able to oxidize the phenolic compound until its complete degradation, achieved within 90 min and 150 min of reaction on ZnO-P-15 and ZnO-C-300. By contrast, a TOC release to the aqueous solution was evidenced during the adsorption period in the dark for both uncleaned ZnO catalysts, before the TOC removal was observed with a kinetic rate constant k'$_{TOC}$ of 0.13 ppm/min and 0.18 ppm/min for ZnO-P-15 and ZnO-C-300 catalysts, respectively (Figure 6b). This TOC release was notably very pronounced in the case of ZnO synthesized via the polyol route, for which no thermal treatment was applied, with a TOC overshoot of 15 ppm at the beginning of the reaction attributed to the release to water of some organic residues from the synthesis, while only 4 ppm of extra TOC was released with the calcined ZnO-C-300 catalyst. This difference in terms of TOC release was in agreement with the higher weight loss recorded on the ZnO-P material and that resulted from the combustion of carbon residues from the synthesis, in comparison to that observed on the calcined ZnO materials.

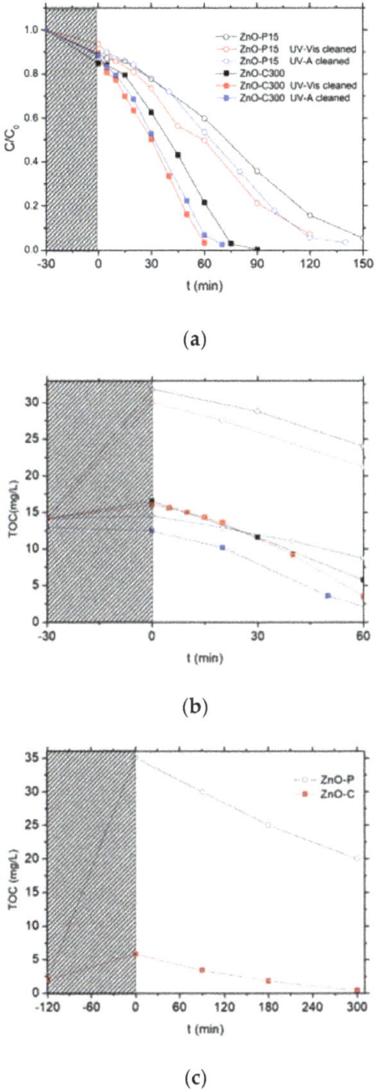

(a)

(b)

(c)

Figure 6. Influence of the photocatalytic pre-cleaning step on (**a**) the relative 4-Cl-phenol concentration and (**b**) the total organic carbon (TOC) concentration evolution upon photocatalysis with ZnO materials. (**c**) Evolution of the TOC concentration during the photocatalytic pre-cleaning step of the as-synthesized ZnO-P and ZnO-C materials under UV-A light in ultrapure Milli-Q water.

The catalyst behavior was strongly affected by applying a photocatalytic cleaning step prior to the chlorophenol removal. This cleaning step consisted in submitting an aqueous suspension of the as-synthesized ZnO catalyst either to simulated solar light (2 h, 25 W/m^2 UV-A irradiance) or to UV-A light (16 h, 60 W/m^2 irradiance). Figure 6c shows in the case of UV-A that those carboned species can be mineralized during the photocatalytic pre-cleaning step of the as-synthesized materials. The volcano-like TOC profile observed has been attributed to the release and the subsequent mineralization of carbon-containing residues coming from the precursors used in the ZnO synthesis, and that remained adsorbed at the catalyst surface or trapped in the bulk of the ZnO crystallites. It confirmed that ZnO synthesized via the polyol method contained larger amounts of carboned residues than its counterparts obtained via the precipitation approach. They were potentially blocking the ZnO active sites, being consequently in competition with the chlorophenol pollutant for the oxidative species, or acting as recombination centers in the ZnO crystallite bulk for the photogenerated charge carriers.

Performing the cleaning procedure under solar light or UV-A light allowed the degradation rates obtained with the ZnO materials to be improved, as can be observed in Figure 6a,b. Both ZnO catalysts cleaned under solar light or UV-A displayed a faster disappearance of chlorophenol than their uncleaned counterparts. It should be noted that performing the pre-cleaning step for 2 h with a 25 W/m^2 UV-A irradiance (i.e., solar light as incident light) already allowed a significant enhancement of the kinetic rate constant for the chlorophenol degradation to be obtained, while by contrast the initial TOC release was only slightly improved. The TOC overshoot was totally suppressed only when the pre-cleaning step was performed under UV-A light for 16 h (60 W/m^2 irradiance).

3.2.2. Photocatalytic Activity of ZnO Materials

Figure 7 shows the evolution with time under irradiation of the 4-Cl-Phenol and TOC concentrations observed on the different ZnO catalysts after the photocatalytic cleaning step under UV-A light has been applied for 2 h. Table 2 shows the corresponding kinetic rate constants for both the 4-Cl-Phenol and the TOC removal.

Irrespective of the cleaned material tested, no significant TOC release was observed at the beginning of the test, confirming the efficiency of the photocatalytic pre-cleaning step. Globally, the ZnO-C materials prepared by precipitation with carbonates displayed the highest activity for both the removal of 4 Cl-phenol and that of TOC when compared to their counterparts prepared via the polyol method or using the carbamates as precipitation agent. Among the carbonate-through precipitation ZnO series, the highest kinetic rate constants of k′$_{Phenol}$ = 0.21 mg/L/min and k′$_{TOC}$ = 0.18 mg/L/min were obtained with the ZnO-C-300 catalyst, for which the complete mineralization of the pollutant was achieved within 70 min of irradiation. This activity improves the results found in the literature for the removal of 4-chloro-phenol with ZnO catalyst under UVA irradiation [36]. Gaya et al. reached the complete 4-Cl-Phenol depletion in 3 h using 2 g/L of ZnO.

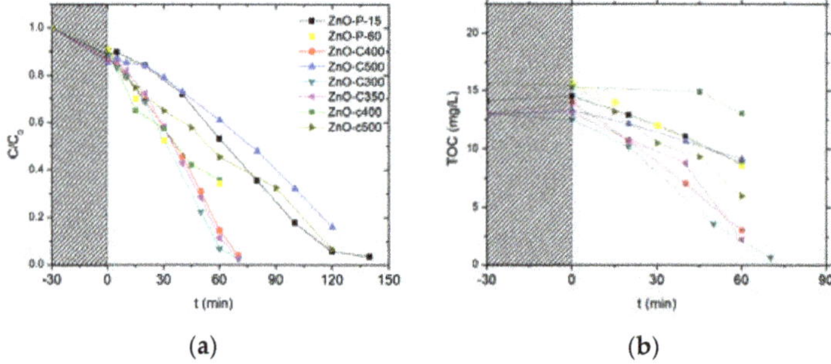

Figure 7. Relative 4-Cl-Phenol concentration (**a**) and TOC concentration (**b**) evolution upon photocatalysis on UV-A light cleaned ZnO materials.

ZnO obtained by the polyol and the carbamate precipitation methods exhibited reduced kinetic constant rates for phenol disappearance and TOC removal, in some cases even two orders of magnitude smaller than that obtained with the most active ZnO. Further, among the ZnO materials prepared with carbonates, ZnO calcined at the highest temperature of 500 °C displayed significantly lower activity compared to that shown by its counterparts calcined in the 300–400 °C range.

We may propose that the ZnO-C-300 photocatalyst takes advantage of a higher specific surface area as well as of a large pore volume and a large mean pore size, that facilitate the access of the 4-chlorophenol reactant and of the reaction intermediates to the ZnO surface sites. By contrast, the high surface area ZnO obtained by the polyol method probably suffered from a very low pore volume and a small mean pore size, while the low surface area of the ZnO-c material prepared with carbamate was assumed to be detrimental to the removal efficiency.

Table 2. Pseudo first-order kinetic rate constant of 4-Cl-Phenol removal and zero order rate constant of TOC removal obtained on UV-A light cleaned ZnO catalysts (with the corresponding linear regression coefficients).

Catalyst	k'_{TOC} (mg/Lmin)/R^2	k'_{Phenol} (min^{-1})/R^2
ZnO-P15	0.097/0.99	0.008/0.98
ZnO-P60	0.130/0.88	0.011/0.94
ZnO-C300	0.180/0.99	0.210/0.99
ZnO-C350	0.173/0.90	0.017/0.99
ZnO-C400	0.180/0.99	0.014/0.99
ZnO-C500	0.071/0.99	0.007/0.97
ZnO-c400	0.030/0.89	0.010/0.93
ZnO-c500	0.153/0.98	0.007/0.98

Surface photocorrosion remains one of the main drawbacks of ZnO for use in water treatment. Photocorrosion consists of the dissolution of the ZnO surface with release of Zn^{2+} to the reaction media, so that the catalyst suffers from deactivation with time under irradiation as well as from intrinsic limitation in terms of reusability. The surface corrosion of ZnO is considered to be induced by the photogenerated holes (h^+) and the overall reaction of the ZnO surface dissolution can be expressed as follows:

$$ZnO + 2h^+ \rightarrow Zn^{2+}_{aqueous} + 0.5O_2 \tag{6}$$

According to previous work, the overall surface photocorrosion process consists of two low-rate steps followed by two high-rate steps as follows [38,39]:

$$O^{2-}_{Surface} + h^+ \rightarrow O^-_{Surface} \tag{7}$$

$$O^-_{Surface} + 3O^{2-} + 3h^+ \rightarrow 2(O - O^{2-}) \tag{8}$$

$$(O - O^{2-}) + 2h^+ \rightarrow O_2 \tag{9}$$

$$2Zn^{2+} \rightarrow 2Zn^{2+}_{aqueous} \tag{10}$$

Therefore, it was of high interest to evaluate the influence of the synthesis method on the stability of the ZnO photocatalysts in terms of Zn^{2+} release to the media, expressed as the percentage of Zn^{2+} released from the ZnO catalysts into the water (Figure 8). In a first approximation, the materials with a higher activity exhibited a higher stability, with a lower release of Zn^{2+} cation to the solution being observed for the most active catalysts and globally a lower release for the ZnO material series prepared by precipitation with carbonates, e.g., a 0.016 fraction in the case of the most active catalyst ZnO-C-300.

The ZnO-C-500 material displayed better stability than its counterparts calcined at 300 °C, but suffered from lower activity under UV-A light. Therefore, based on the kinetic rate constants derived from both the chlorophenol disappearance and the TOC evolution curves, as well as on the level of Zn^{2+} release to the media, this fast screening allowed selecting the ZnO-C-300 material obtained via the precipitation method with carbonates and calcined at 300 °C, for implementing subsequently the solar light photon-assisted preparation of the Cu-ZnO hybrid catalysts in the next section.

However, we would like to point out that the surface photocorrosion of ZnO materials remained a key issue for using such materials as photocatalysts in water treatment. Indeed, a strong loss of activity was observed when performing sequential runs on the ZnO-C-300 photocatalyst. Using the TOC conversion achieved after 75 min of test as an indicator (i.e., the duration necessary for achieving full TOC conversion in the run#1), Table 3 evidences that the surface photocorrosion was accompanied by an important loss of activity with sequential runs.

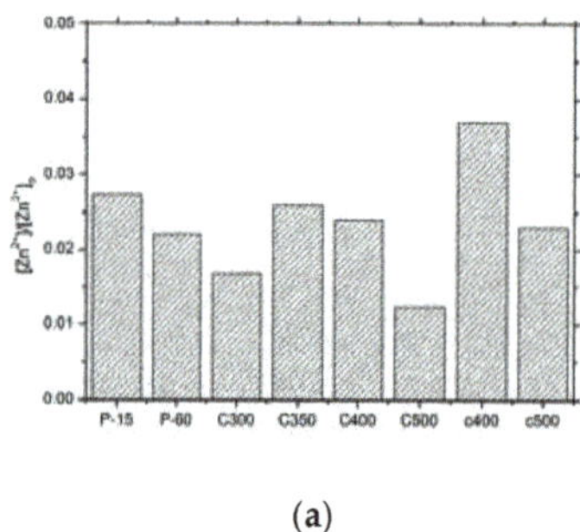

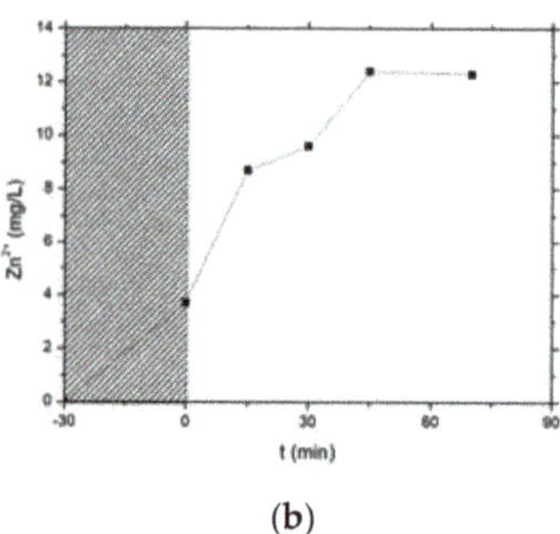

(a)   (b)

Figure 8. Fraction of Zn^{2+} released into the water (**a**) at the end of the photocatalytic run with ZnO materials and (**b**) as a function of time under irradiation in the case of the ZnO-C-400 catalyst.

Table 3. TOC conversion after several sequential photocatalytic runs with the ZnO-C300 catalyst.

Run	X_{TOC} (%)
First	100
Second	88
Third	61

3.3. Preparation of Cu-ZnO Catalysts by the Photon-Assisted Synthesis Method

The photon-assisted preparation of the Cu-ZnO catalysts was performed using the ZnO-C-300 material as semi-conductor support, since the first part of this work evidenced that it displayed the highest photoactivity in terms of removal of both chlorophenol and TOC, and also showed a high photo-stability in water. The photon-assisted synthesis method takes advantage of the redox photo-activity of the ZnO semiconductor under solar light for allowing the reduction of the metal ions adsorbed at the host surface by the photogenerated electrons from the conduction band of the irradiated semi-conductor. Both $Cu(acac)_2$ and $Cu(NO_3)_2$ metallic precursors were used, and Figure 9 depicts the disappearance curves of both precursors during the photon-assisted synthesis in the presence of the ZnO-C-300 catalyst, derived from the time-evolution of the UV-vis absorbance spectra.

First, photolysis of the copper precursors under solar light in our experimental conditions could be disregarded, since no changes of the UV-vis spectra were observed irrespective of the precursor used (not shown). The evolution with time of the C/C_0 relative concentrations evidenced that the photodeposition occurred on the ZnO support, and further that the copper precursor nature influenced the kinetics of the photodeposition process at the surface of the irradiated ZnO support. However, it has to be noted that time-monitoring of the photo-deposition process was more difficult using the nitrate precursor than its acetylacetonate counterpart. Indeed, the ability of semi-conductor photocatalysis to perform nitrate reduction and nitrate oxidation in water led to the production first of nitrite as initial intermediate (very unstable and easily oxidized back to nitrate) and further higher oxidation state products such as ammonia. The UV-vis signature of those N-compounds is known to overlap around similar absorption wavelengths, so that the photodeposition was monitored in a first approximation by following the disappearance of the low intensity absorption peak at $\lambda = 800$ nm related to the Cu^{2+} concentration, which resulted in less accurate time-monitoring. However, ICP-OES analysis revealed that the Cu-ZnO catalyst had a Cu content of 9.7 wt %, in good agreement with the targeted theoretical amount of 10 wt %. By contrast, only a Cu content of 7 wt % was obtained in the case of the Cu(acac)$_2$ precursor. This difference was attributed to a possible release of non-steadily anchored Cu species during the washing of the Cu-ZnO materials directly after the photon-assisted synthesis.

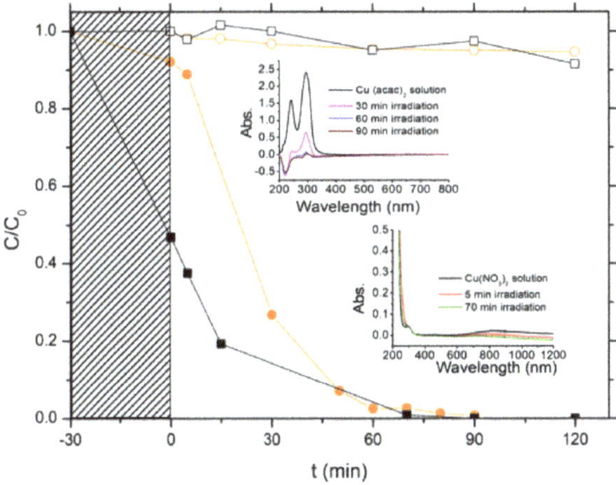

Figure 9. Disappearance curves of (■) the Cu(NO$_3$)$_2$ and (•) the Cu(acac)$_2$ precursors during the photon-assisted synthesis in the presence of the ZnO-C-300 catalyst. Blank photolysis experiments in the absence of ZnO photocatalyst for (□) Cu(NO$_3$)$_2$ and (○) Cu(acac)$_2$ precursors. Insert: Time-evolution of the UV-vis absorbance spectra for both precursors (selected analysis times are reported for not overloading the UV-vis absorbance graphs).

3.4. Characterization of Cu-ZnO Catalysts

Figure 10 shows the XRD patterns of both Cu-ZnO catalysts. In the case of the Cu(acac)$_2$ precursor, besides the diffraction peaks of ZnO crystallized in the hexagonal wurtzite structure, additional diffraction peaks were observed at 29.6°, 42.2°, 61.3° and 73.4°, and attributed to the diffraction of (110), (111), (220) and (311) planes of cubic Cu$^{(I)}_2$O nanoparticles with an average crystallite size of 65 nm (JCPDS Card 00-005-0667). By contrast, in the case of Cu(NO$_3$)$_2$ precursor, additional diffraction peaks and shoulders have been recorded at 2θ = 38.8°, 54.0° and 58.0°, and attributed to the diffraction of (111), (020) and (202) planes of monoclinic Cu$^{(II)}$O nanoparticles with an average crystallite size of 10 nm (JCPDS Card 01-089-5895). Interestingly, the choice of the photon-assisted synthesis/deposition parameters allowed selective driving of the oxidation state of the Cu nanoparticles synthesized on the ZnO support towards either cuprous or cupric oxides, so that the prepared catalysts could be considered as Cu$^{(I)}_2$O-ZnO or Cu$^{(II)}$O-ZnO. This suggests the possibility of controlling the oxidation state of the Cu nanoparticles in the Cu-ZnO while implementing a preparation procedure at room temperature without applying any final thermal treatment.

In the case of the Cu nitrate precursor, the reduction of adsorbed Cu^{2+} ions by the photogenerated electrons would first give Cu^0, which might be further re-oxidized by the OH radicals formed via the oxidation of adsorbed water by the photogenerated holes from the valence band, or directly by the holes. Cu^0 was also suggested to undergo reoxidation by O$_2$ into Cu^{2+} [40], the Cu^{2+}/Cu^0 couple being then considered to act as an electron mediator from the conduction band to O$_2$. In the case of the acetylacetonate precursor, the direct Cu(acac)$_2$ reduction into Cu^0 was reported to be strongly unfavored compared to that of Cu^{2+} [40]. By contrast, the adsorbed acetylacetonate precursor was proposed to be first oxidized by OH radicals formed via the oxidation of adsorbed water by the photogenerated holes from the valence band, or directly by the holes. Subsequently, the ligand oxidation would generate adsorbed Cu^{2+} ions that can be subsequently reduced by the photogenerated electrons.

Further complementary research is being performed for understanding the mechanisms involved in the selective formation of Cu^{2+} or Cu^+ species at the surface of the ZnO support and for shedding light on the main synthesis parameters that enable driving of the oxidation state of the Cu within the Cu-ZnO material.

It should be noted that performing the photodeposition process on the ZnO support led to an increase in the mean crystallite site of the semi-conductor support from 10 nm to 20 nm in the case of the copper nitrate precursor. By contrast, no significant change was observed in the case of the acetylacetonate precursor, with a mean crystallite size of 13 nm being obtained.

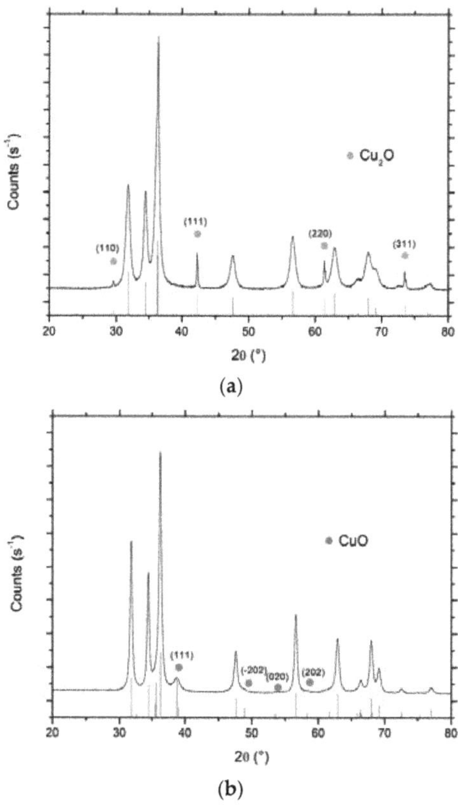

Figure 10. XRD patterns of the Cu_xO-ZnO materials synthesized by photodeposition using (a) $Cu(acac)_2$ and (b) $Cu(NO_3)_2$ as metallic precursor. Main reflexes from JCPDS cards No 01-089-5895 and 00-005-0667 for CuO and Cu_2O phases, respectively. The main reflexes from JCPDS card No 00-036-1451 for the ZnO support are shown in blue.

The main physico-chemical properties of the Cu_xO/ZnO materials are reported in Table 1. It evidences that the synthesis of the CuO-ZnO and Cu_2O-ZnO composite materials by photodeposition resulted in specific surface area and pore volume changes, and the composite materials displayed specific surface areas of 33 m^2/g and 95 m^2/g, respectively, with pore volumes of 0.24 cm^3/g and 0.68 cm^3/g. Figure 11 shows that the photon-assisted synthesis of Cu_xO did not modify either isotherm or hysteresis types of type II or H3, respectively, while it slightly influenced the pore size distribution profiles. In addition, the SEM images shown in Figure 12 indicate that the overall morphology of the ZnO-based materials was not modified by the photon-assisted synthesis process.

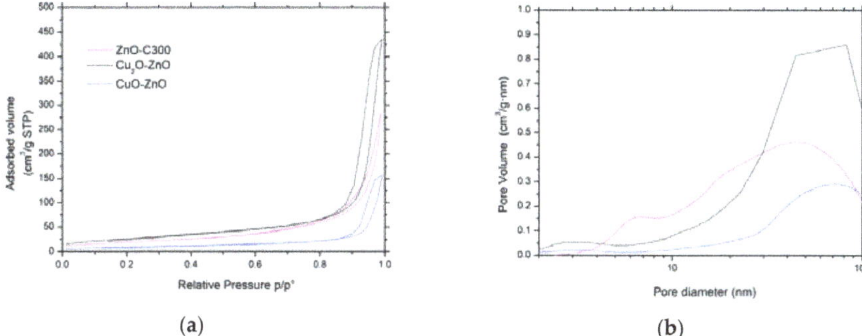

Figure 11. (a) N_2 adsorption-desorption isotherms and (b) pore size distributions of both Cu_xO-ZnO catalysts and bare ZnO material as reference.

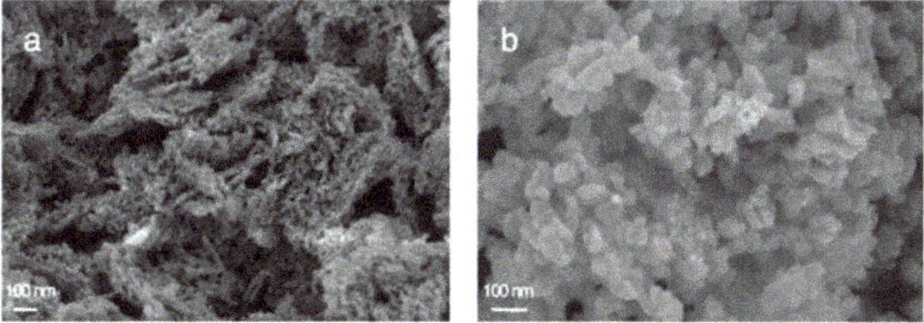

Figure 12. SEM images of (a) Cu_2O-ZnO and (b) CuO-ZnO catalysts.

Figure 13 shows the TEM images of the Cu-ZnO materials obtained by the photon-assisted synthesis method with both copper precursors. Unfortunately, the similarities of the interplanar spacings corresponding to the main planes of the ZnO material and of both $Cu^{(II)}O$ and $Cu_2^{(I)}O$ crystallites did not allow the specific identification and location of the different phases within the Cu-ZnO materials that consisted of very entangled oxide crystallites in close contact. Indeed, the (101) and (100) planes of ZnO have interplanar distances of 0.25 nm and 0.28 nm (JCPDS No.00-036-1451), while the (111) and (002) planes of the $Cu^{(II)}O$ phase have 0.23 nm and 0.25 nm interplanar distances, respectively (JCPDS No. 01-089-5895) [41], and the (111) planes of the $Cu_2^{(I)}O$ crystallites have an interplanar distance of 0.25 nm (JCPDS No. 00-005-0667) [42]. However, the presence of copper has been confirmed by performing overall EDS analysis on both samples, with a Cu content of 7.4 wt % and 9.5 wt % for the CuO-ZnO and the Cu_2O-ZnO materials, respectively, in good agreement with the ICP-OES data.

It is worth noting that irrespective of the copper precursor, the prepared material can be considered as a Cu-ZnO composite catalyst, since both Cu_2O and CuO crystallites do not exhibit significantly smaller mean sizes than the ZnO crystallites, rather than as a ZnO supported Cu catalyst that would usually consist of smaller size Cu-based nanoparticles dispersed on the ZnO support. Considering the exclusive presence of one single chemical state for the crystalline Cu-based crystallites in the samples as shown by XRD—keeping in mind the detection limit of the XRD measurement—the Cu-ZnO materials can thus be considered as $Cu^{(II)}O$-ZnO and $Cu_2^{(I)}O$-ZnO composite catalysts, although the assignment of both transition metal oxide phases to specific crystallites in TEM images was not possible.

In addition, we cannot rule out that in addition to larger crystallites, some Cu species might remain highly dispersed on ZnO, thanks to a strong interaction between Cu and the amphoteric ZnO material that seems to protect the Cu nanoparticles from sintering, consequently providing enhanced interphase contact [43].

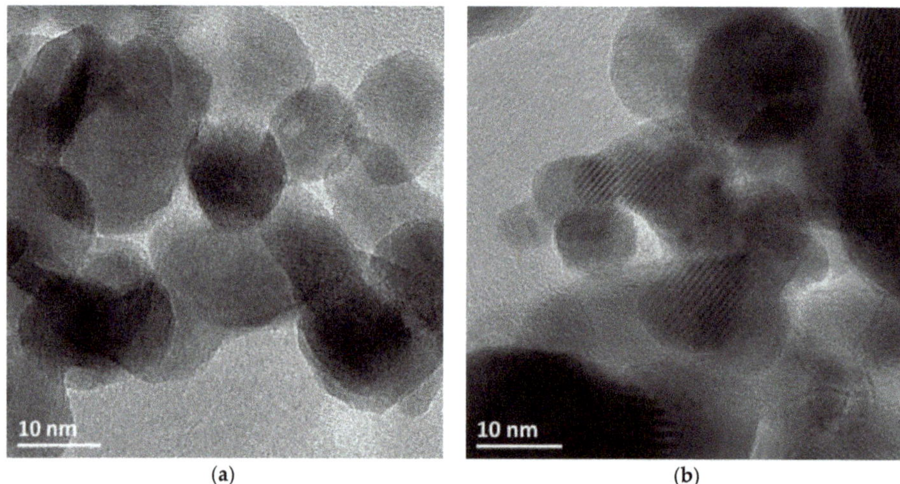

(a)   (b)

Figure 13. TEM images of the Cu-ZnO materials synthesized by photodeposition using (**a**) $Cu(acac)_2$ and (**b**) $Cu(NO_3)_2$ as metallic precursor. An example of STEM imaging is shown as Figure S2.

4. Conclusions

Cu_xO-ZnO composite catalysts with a control of the chemical state of the copper oxide phase were prepared at room temperature using the solar light-induced redox photoactivity of the ZnO semiconductor support. The preparation of $Cu_2^{(I)}O$-ZnO and $Cu^{(II)}O$-ZnO composite catalysts was achieved using $Cu(acac)_2$ in THF-water and $Cu(NO_3)_2$ in water as metallic precursor, respectively.

The photoactive ZnO host material was prepared through the precipitation method with carbonate as precipitation agent, and its superior activity compared to ZnO synthesized by other methods was evaluated by taking the photocatalytic degradation of the 4-chlorophenol compound in water under simulated solar light as a model reaction.

Research is ongoing for understanding the key parameters driving the selective synthesis of the Cu_xO phase within the ZnO-based catalysts. The activity of Cu_xO-ZnO catalysts will be investigated in high interest reactions in the field of biomass conversion. Besides thermal catalysis applications, this work has opened a new route for the facile synthesis of Cu_2O-ZnO heterojunction photocatalysts, which could take advantage under solar light of the heterojunction built between the *p*-type semi-conductor Cu_2O with direct band gap of about 2.17 eV and the ZnO semiconductor phase.

Supplementary Materials: The following are available online at http://www.mdpi.com/1996-1944/11/11/2260/s1. Figure S1: (a) Absorbance spectra of selected ZnO and Cu_xO-ZnO photocatalysts and (b) the corresponding $(\alpha h\upsilon)^{1/2}$ vs. $h\upsilon$ plots used to estimate the band gap of the ZnO materials using the K-M function $F(R)$. Figure S2: Mapping STEM imaging recorded on CuO-ZnO composite material prepared from the Cu nitrate precursor: (red) Zn K, (blue) Cu K, (green) O K and (overlay).

Author Contributions: Experimental works, M.B. and P.G.M.; Data analysis, M.B., P.G.M., A.M.R. and N.K.; Writing—Original Draft Preparation, P.G.M. and N.K.; Visualization, P.G.M. and M.B.; Supervision, N.K. and A.M.R.; Project Administration, N.K.; Funding Acquisition, M.B., N.K. and A.M.R.

Funding: The authors gratefully acknowledge that this work was financially supported by a grant from the National Center of Science (NCN) in Krakow (Poland) (2016/22/E/ST4/00550). This research was co-funded by the Erasmus+ program of the European Union.

Acknowledgments: T. Dintzer (ICPEES, Strasbourg) and D. Ihiawakrim (IPCMS, Strasbourg) are thanked for performing SEM and TEM characterization, respectively.

Conflicts of Interest: The authors declare no conflict of interest.

References

1. Kołodziejczak-Radzimska, A.; Jesionowski, T. Zinc oxide from synthesis to application: A review. *Materials* **2014**, *7*, 2833–2881. [CrossRef] [PubMed]

2. Segets, D.; Gradl, J.; Taylor, R.K.; Vassilev, V.; Peukert, W. Analysis of optical absorbance spectra for the determination of ZnO nanoparticle size distribution, solubility, and surface energy. *ACS Nano* **2009**, *3*, 1703–1710. [CrossRef] [PubMed]

3. Chaari, M.; Matoussi, A. Electrical conduction and dielectric studies of ZnO pellets. *Phys. B Condens. Matter* **2012**, *407*, 3441–3447. [CrossRef]

4. Bacaksiz, E.; Parlak, M.; Tomakin, M.; Özçelik, A.; Karakız, M.; Altunbaş, M. The effects of zinc nitrate, zinc acetate and zinc chloride precursors on investigation of structural and optical properties of ZnO thin films. *J. Alloys Compd.* **2008**, *466*, 447–450. [CrossRef]

5. Wang, Z.L. Splendid one-dimensional nanostructures of zinc oxide: A new nanomaterial family for nanotechnology. *ACS Nano* **2008**, *2*, 1987–1992. [CrossRef] [PubMed]

6. Wang, J.; Cao, J.; Fang, B.; Lu, P.; Deng, S.; Wang, H. Synthesis and characterization of multipod, flower-like, and shuttle-like ZnO frameworks in ionic liquids. *Mater. Lett.* **2005**, *59*, 1405–1408. [CrossRef]

7. Sudha, D.; Sivakumar, P. Review on the photocatalytic activity of various composite catalysts. *Chem. Eng. Process. Process Intensif.* **2015**, *97*, 112–133. [CrossRef]

8. Janotti, A.; Van de Walle, C.G. Fundamentals of zinc oxide as a semiconductor. *Rep. Prog. Phys.* **2009**, *72*, 126501. [CrossRef]

9. Reynolds, D.C.; Look, D.C.; Jogai, B.; Litton, C.W.; Cantwell, G.; Harsch, W.C. Valence-band ordering in ZnO. *Phys. Rev. B* **1999**, *60*, 2340–2344. [CrossRef]

10. Chen, Y.; Bagnall, D.M.; Koh, H.; Park, K.; Hiraga, K.; Zhu, Z.; Yao, T. Plasma assisted molecular beam epitaxy of ZnO on c-plane sapphire: Growth and characterization. *J. Appl. Phys.* **1998**, *84*, 3912–3918. [CrossRef]

11. Udom, I.; Ram, M.K.; Stefanakos, E.K.; Hepp, A.F.; Goswami, D.Y. One dimensional-ZnO nanostructures: Synthesis, properties and environmental applications. *Mater. Sci. Semicond. Process.* **2013**, *16*, 2070–2083. [CrossRef]

12. Hernández-Carrillo, M.A.; Torres-Ricárdez, R.; García-Mendoza, M.F.; Ramírez-Morales, E.; Rojas-Blanco, L.; Díaz-Flores, L.L.; Sepúlveda-Palacios, G.E.; Paraguay-Delgado, F.; Pérez-Hernández, G. Eu-modified ZnO nanoparticles for applications in photocatalysis. *Catal. Today* **2018**. [CrossRef]

13. Machín, A.; Cotto, M.; Duconge, J.; Arango, J.C.; Morant, C.; Pinilla, S.; Soto-Vázquez, L.; Resto, E.; Márquez, F. Hydrogen production via water splitting using different Au@ZnO catalysts under UV–vis irradiation. *J. Photochem. Photobiol. A Chem.* **2018**, *353*, 385–394. [CrossRef]

14. Fkiri, A.; Santacruz, M.R.; Mezni, A.; Smiri, L.S.; Keller, V.; Keller, N. One-pot synthesis of lightly doped $Zn_{1-x}Cu_xO$ and $Au–Zn_{1-x}Cu_xO$ with solar light photocatalytic activity in liquid phase. *Environ. Sci. Pollut. Res.* **2017**, *24*, 15622–15633. [CrossRef] [PubMed]

15. Liu, Y.; Zhong, M.; Shan, G.; Li, Y.; Huang, B.; Yang, G. Biocompatible ZnO/Au nanocomposites for ultrasensitive DNA detection using resonance Raman scattering. *J. Phys. Chem. B* **2008**, *112*, 6484–6489. [CrossRef] [PubMed]

16. Udawatte, N.; Lee, M.; Kim, J.; Lee, D. Well-defined Au/ZnO nanoparticle composites exhibiting enhanced photocatalytic activities. *ACS Appl. Mater. Interfaces* **2011**, *3*, 4531–4538. [CrossRef] [PubMed]

17. Li, H.; Liu, E.; Chan, F.Y.F.; Lu, Z.; Chen, R. Fabrication of ordered flower-like ZnO nanostructures by a microwave and ultrasonic combined technique and their enhanced photocatalytic activity. *Mater. Lett.* **2011**, *65*, 3440–3443. [CrossRef]

18. Li, P.; Wei, Z.; Wu, T.; Peng, Q.; Li, Y. Au-ZnO hybrid nanopyramids and their photocatalytic properties. *J. Am. Chem. Soc.* **2011**, *133*, 5660–5663. [CrossRef] [PubMed]

19. Chen, P.; Lee, G.; Anandan, S.; Wu, J.J. Synthesis of ZnO and Au tethered ZnO pyramid-like microflower for photocatalytic degradation of orange II. *Mater. Sci. Eng. B* **2012**, *177*, 190–196. [CrossRef]

20. Sun, Q.; Men, Y.; Wang, J.; Chai, S.; Song, Q. Support effect of Ag/ZnO catalysts for partial oxidation of methanol. *Inorg. Chem. Commun.* **2018**, *92*, 51–54. [CrossRef]

21. Grunwaldt, J.D.; Molenbroek, A.M.; Topsøe, N.Y.; Topsøe, H.; Clausen, B.S. In situ investigations of structural changes in Cu/ZnO catalysts. *J. Catal.* **2000**, *194*, 452–460. [CrossRef]

22. Ichikawa, M. Catalysis by supported metal crystallites from carbonyl clusters. I. Catalytic methanol synthesis under mild conditions over supported rhodium, platinum, and iridium crystallites prepared from Rh, Pt, and Ir carbonyl cluster compounds deposited on ZnO and MgO. *Bull. Chem. Soc. Jpn.* **1978**, *51*, 2268–2272. [CrossRef]

23. Cubeiro, M.L.; Fierro, J.L.G. Selective production of hydrogen by partial oxidation of methanol over ZnO-supported palladium catalysts. *J. Catal.* **1998**, *179*, 150–162. [CrossRef]

24. Zheng, L.; Li, X.; Du, W.; Shi, D.; Ning, W.; Lu, X.; Hou, Z. Metal-organic framework derived Cu/ZnO catalysts for continuous hydrogenolysis of glycerol. *Appl. Catal. B Environ.* **2017**, *203*, 146–153. [CrossRef]

25. Alba-Rubio, A.C.; Santamaría-González, J.; Mérida-Robles, J.M.; Moreno-Tost, R.; Martín-Alonso, D.; Jiménez-López, A.; Maireles-Torres, P. Heterogeneous transesterification processes by using CaO supported on zinc oxide as basic catalysts. *Catal. Today* **2010**, *149*, 281–287. [CrossRef]

26. Zhu, Y.; Kong, X.; Zheng, H.; Ding, G.; Zhu, Y.; Li, Y. Efficient synthesis of 2,5-dihydroxymethylfuran and 2,5-dimethylfuran from 5-hydroxymethylfurfural using mineral-derived Cu catalysts as versatile catalysts. *Catal. Sci. Technol.* **2015**, *5*, 4208–4217. [CrossRef]

27. Llorca, J.; de la Piscina, P.R.; Dalmon, J.; Sales, J.; Homs, N. CO-free hydrogen from steam-reforming of bioethanol over ZnO-supported cobalt catalysts: Effect of the metallic precursor. *Appl. Catal. B Environ.* **2003**, *43*, 355–369. [CrossRef]

28. Homs, N.; Llorca, J.; de la Piscina, P.R. Low-temperature steam-reforming of ethanol over ZnO-supported Ni and Cu catalysts: The effect of nickel and copper addition to ZnO-supported cobalt-based catalysts. *Catal. Today* **2006**, *116*, 361–366. [CrossRef]

29. Ammari, F.; Lamotte, J.; Touroude, R. An emergent catalytic material: Pt/ZnO catalyst for selective hydrogenation of crotonaldehyde. *J. Catal.* **2004**, *221*, 32–42. [CrossRef]

30. Spencer, M.S. The role of zinc oxide in Cu/ZnO catalysts for methanol synthesis and the water gas shift reaction. *Top. Catal.* **1999**, *8*, 259. [CrossRef]

31. Pinna, F. Supported metal catalysts preparation. *Catal. Today* **1998**, *41*, 129–137. [CrossRef]

32. Liu, X.; Liu, M.H.; Luo, Y.C.; Mou, C.Y.; Lin, S.D.; Cheng, H.; Chen, J.M.; Lee, J.F.; Lin, T.S. Strong metal-support interactions between gold nanoparticles and ZnO nanorods in CO oxidation. *J. Am. Chem. Soc.* **2012**, *134*, 10251. [CrossRef] [PubMed]

33. Le Valant, A.; Comminges, C.; Tisseraud, C.; Canaff, C.; Pinard, L.; Pouilloux, Y. The Cu–ZnO synergy in methanol synthesis from CO_2, Part 1: Origin of active site explained by experimental studies and a sphere contact quantification model on Cu + ZnO mechanical mixtures. *J. Catal.* **2015**, *324*, 41–49. [CrossRef]

34. Collins, S.S.E.; Cittadini, M.; Pecharromán, C.; Martucci, A.; Mulvaney, P. Hydrogen spillover between single gold nanorods and metal oxide supports: A surface plasmon spectroscopy study. *ACS Nano* **2015**, *9*, 7846. [CrossRef] [PubMed]

35. Lee, K.M.; Lai, C.W.; Ngai, K.S.; Juan, J.C. Recent developments of zinc oxide based photocatalyst in water treatment technology: A review. *Water Res.* **2016**, *88*, 428–448. [CrossRef] [PubMed]

36. Gaya, U.I.; Abdullah, A.H.; Zainal, Z.; Hussein, M.Z. Photocatalytic treatment of 4-chlorophenol in aqueous ZnO suspensions: Intermediates, influence of dosage and inorganic anions. *J. Hazard. Mater.* **2009**, *168*, 57–63. [CrossRef] [PubMed]

37. Eslami, A.; Hashemi, M.; Ghanbari, F. Degradation of 4-chlorophenol using catalyzed peroxymonosulfate with nano-MnO_2/UV irradiation: Toxicity assessment and evaluation for industrial wastewater treatment. *J. Clean. Prod.* **2018**, *195*, 1389–1397. [CrossRef]

38. Ma, X.; Li, H.; Liu, T.; Du, S.; Qiang, Q.; Wang, Y.; Yin, S.; Sato, T. Comparison of photocatalytic reaction-induced selective corrosion with photocorrosion: Impact on morphology and stability of Ag-ZnO. *Appl. Catal. B Environ.* **2017**, *201*, 348–358. [CrossRef]

39. Han, C.; Yang, M.; Weng, B.; Xu, Y. Improving the photocatalytic activity and anti-photocorrosion of semiconductor ZnO by coupling with versatile carbon. *Phys. Chem. Chem. Phys.* **2014**, *16*, 16891–16903. [CrossRef] [PubMed]

40. Naya, S.I.; Tanaka, M.; Kimura, K.; Tada, H. Visible-light-driven copper acetylacetonate decomposition by BiVO$_4$. *Langmuir* **2011**, *27*, 10334–10339. [CrossRef] [PubMed]

41. Liu, Y.; Zhong, L.; Peng, Z.; Song, Y.; Chen, W. Field emission properties of one-dimensional single CuO nanoneedle by in situ microscopy. *J. Mater. Sci.* **2010**, *45*, 3791–3796. [CrossRef]

42. Kim, M.H.; Lim, B.; Lee, E.; Xia, Y. Polyol synthesis of Cu$_2$O nanoparticles: Use of chloride to promote the formation of a cubic morphology. *J. Mater. Chem.* **2008**, *18*, 4069–4073. [CrossRef]

43. Chen, S.; Wojcieszak, R.; Dumeignil, F.; Marceau, E.; Royer, S. How Catalysts and experimental conditions determine the selective hydroconversion of furfural and 5-hydroxymethylfurfural. *Chem. Rev.* **2018**. [CrossRef] [PubMed]

© 2018 by the authors. Licensee MDPI, Basel, Switzerland. This article is an open access article distributed under the terms and conditions of the Creative Commons Attribution (CC BY) license (http://creativecommons.org/licenses/by/4.0/).

 *materials*

Article

Photocatalytic Activity of Nanotubular TiO$_2$ Films Obtained by Anodic Oxidation: A Comparison in Gas and Liquid Phase

Beatriz Eugenia Sanabria Arenas [1], Alberto Strini [2], Luca Schiavi [2], Andrea Li Bassi [3], Valeria Russo [3], Barbara Del Curto [1], Maria Vittoria Diamanti [1,*] and MariaPia Pedeferri [1]

[1] Department of Chemistry, Materials and Chemical Engineering "G. Natta" Politecnico di Milano, Via Mancinelli 7, Milan 20131, Italy; beatrizeugenia.sanabria@polimi.it (B.E.S.); barbara.delcurto@polimi.it (B.D.C.); mariapia.pedeferri@polimi.it (M.P.)
[2] ITC-CNR, Construction Technologies Institute, Viale Lombardia 49, San Giuliano Milanese, Milan 20098, Italy; alberto.strini@itc.cnr.it (A.S.); luca.schiavi@itc.cnr.it (L.S.)
[3] Department of Energy, Politecnico di Milano, Via Ponzio 34/3, Milan 20133, Italy; andrea.libassi@polimi.it (A.L.B.); valeria.russo@polimi.it (V.R.)
* Correspondence: mariavittoria.diamanti@polimi.it; Tel.: +39-02-2399-3137

Received: 20 February 2018; Accepted: 20 March 2018; Published: 24 March 2018

Abstract: The availability of immobilized nanostructured photocatalysts is of great importance in the purification of both polluted air and liquids (e.g., industrial wastewaters). Metal-supported titanium dioxide films with nanotubular morphology and good photocatalytic efficiency in both environments can be produced by anodic oxidation, which avoids release of nanoscale materials in the environment. Here we evaluate the effect of different anodizing procedures on the photocatalytic activity of TiO$_2$ nanostructures in gas and liquid phases, in order to identify the most efficient and robust technique for the production of TiO$_2$ layers with different morphologies and high photocatalytic activity in both phases. Rhodamine B and toluene were used as model pollutants in the two media, respectively. It was found that the role of the anodizing electrolyte is particularly crucial, as it provides substantial differences in the oxide specific surface area: nanotubular structures show remarkably different activities, especially in gas phase degradation reactions, and within nanotubular structures, those produced by organic electrolytes lead to better photocatalytic activity in both conditions tested.

Keywords: nanostructured materials; titanium dioxide; anodizing; photocatalysis; toluene; rhodamine B

---

1. Introduction

Heterogeneous photocatalysis using semiconductors is an inexpensive and environmentally friendly treatment technology for degrading organic and inorganic pollutants in gas and liquid phase, which presents several advantages, such as minimal equipment and reuse of catalyst [1,2]. The number of photocatalytic studies reporting liquid phase reactions is far higher than those related to gas phase; however, this tendency has started to change in recent years, due to different applications of this technology in indoor environments like office buildings, factories, aircraft and spacecraft [3]. In such environments, the biggest concerns are related not only to pollution coming from outdoor activities—such as NO$_x$ produced by vehicles and heaters combustion processes—but also from inside: paints, varnishes, detergents and furniture are all sources of volatile organic compounds, such as benzene, toluene and formaldehyde, which pose health risks to the inhabitants, as stated by WHO—World Health Organization guidelines for indoor quality.

On the one hand, a strong correlation between chemical nature of pollutant, catalyst surface characteristics, concentration of water vapour and obtained photoactivity has been reported in gas-solid photoreactions [4]. Einaga et al. [5] studied the degradation of toluene at different concentrations and water vapour content and reported a dependence between the initial concentration of toluene and the type of intermediates formed, as well as the necessity of water vapour in the system for complete mineralization of the intermediates to CO_2 and CO. Nonetheless, one of the main challenges in this type of photocatalysis is to avoid the deactivation of the catalyst due to the accumulation of decomposition products at the surface of the substrate [6]. On the other hand, in liquid phase the number of variables is greater than in gas phase, and therefore the complexity. Surface composition, surface area, preparation procedures, concentration of catalyst, pH, concentration and chemical structure of the pollutant, partial pressure of oxygen and diffusion rate, are some of the parameters that can influence the photocatalysis of solid-liquid reactions [7]. Yet, photocatalysis has proved to be an effective technique for both water and air purification purposes, and several materials are available for the task, with different characteristics, depending on the specific application: from TiO_2 to ZnO, from WO_3, to SnO_2. These materials are available in form of nanoparticles, nanowires or nanotubes, and modifications of chemical compositions can enhance the activity of these materials making them suitable also for visible light applications [8,9].

Most of the studies about photocatalytic degradation with TiO_2 in both liquid and gas phase have been performed using commercial available nanopowders like AEROXIDE® TiO_2 P25, previously known as Degussa P25, which is either chosen as model photocatalyst, or used as comparison due to the large amount of data available on this material [10–15]. The dispersion of nanoparticles in liquid phase makes it necessary to add one or more recovery steps after photocatalysis, increasing the process costs and complexity, especially at industrial scale [16–18]. Other problems regarding the use of nanopowders are agglomeration in aqueous suspensions, which decreases photocatalytic activity, and light scattering caused by the same nanoparticles, which limits light absorption in large reactors [2].

Immobilized nanoparticles have been proposed as solution to this problem. This can be done over different materials [3] but glass has been by far the most used because it allows the penetration of light. However, the impossibility to use organic matrices such as common paints due to their photocatalytic destruction [1], lower photocatalytic activity due to decrease in surface area and light absorption [19], as well as limited mass diffusion of reactants within an immobilizing medium—be it organic or inorganic—are some of the disadvantages found by using this methodology, requiring the use of more resistant binders, such as perfluorinated ones [20,21].

Photocatalytic nanoparticles can also be immobilized onto inorganic substrates by high-temperature sintering, as example after deposition from a slurry [18] or by a suitable film-making process such as screen-printing [22]. However, TiO_2 sintering is strongly limited by the anatase phase thermal instability, as it converts to the far less active rutile above 500 °C, preventing to reach the higher temperatures needed to ensure an optimal film strength. Improvements can be obtained by doping the anatase with suitable metal cations such as Zr^{4+} [23] that act as phase stabilizers. This can be done during the nanoparticle synthesis or also with undoped anatase during the sintering itself, but the process temperature is anyway limited to ~800 °C in order to avoid a significant degradation of the photocatalytic activity [24].

Anodic oxidation is a powerful technique that allows the production of various types of oxides on the surface of titanium [25], including highly ordered nanotubular oxides with comparable or higher photocatalytic activity than nanoparticles [26,27], and strong adherence to a metallic substrate [19]. Such structures are produced by anodizing the metal in solutions containing fluorides (or, more broadly, halogens), which dissolve the oxide locally while it is growing, allowing the formation of self-organized nanopores or nanotubes. This technique allows to tune very important parameters such as pore diameter, tube length, oxide thickness, wall smoothness, and porosity which affects physical, electronic, and chemical properties that finally have repercussions in the photocatalytic efficiency [28–34]. After nanotubes anodizing it is mandatory to anneal the sample at temperatures between 300 and 500 °C to induce oxide crystallization to anatase phase, which is acknowledged to be more efficient than rutile, owing to its higher charge carrier mobility [16]. On the other hand, anodizing at high voltages in sulfuric acid or other electrolytes allows direct crystallization of the oxide during its growth, on account of the onset of anodic spark deposition conditions: the higher the voltage, the higher anatase and rutile contents in the oxide, accompanied by a sub-micrometric porosity [35].

The lack of reproducibility, however, is still the main problem for this new technology to find a widespread use in industrial applications. In order to overcome these difficulties, clear correlations between anodizing parameters, material characteristics and obtained photoactivity should be thoroughly investigated, According to Zhuiykov [36] the morphology, crystal structure, surface stoichiometry, catalytic activity and chemical composition can be affected by the process used to produce the nanostructures and it explains why it is so difficult to reproduce the results obtained by other researchers. In other words, a particular fabrication process can affect the pathway that the molecules follow during degradation.

For these reasons, this work proposes a comparison among several different anodizing treatments featuring different process parameters, which lead to different oxide morphology and crystallinity. All of them were tested in the photocatalytic degradation of liquid pollutants, in particular the organic dye rhodamine B (RhB), and in gas phase photocatalytic degradation of toluene. This is meant to provide a reliable comparison among different anodizing techniques and procedures, as all tests are performed in a homogeneous way, thus solving the issue of comparing results provided by different research studies. Three categories of anodic oxides were chosen: anodic spark deposition (ASD) oxides, nanotubes produced in aqueous solution and nanotubes obtained in organic solution. The final aim of this work is to present a robust method to obtain immobilized TiO_2 nanostructures by anodic oxidation, and to allow for the most suitable anodizing choice as a function of the desired application, be it gas or liquid phase, taking into account method robustness and ease of reproduction.

2. Materials and Methods

2.1. Chemicals and TiO$_2$ Nanostructures Synthesis

All tests were performed on commercial purity titanium sheets (grade 2 ASTM), employing NaF (Fluka), ethylene glycol (Fluka) and H_2SO_4 (Merck) for the electrolytes, and studying the photocatalytic degradation of Rhodamine B (Sigma) and toluene (diluted standard cylinder in N_2, SIAD, Italy) in presence of the anodized titanium. All the reagents used were analytical grade.

Anodic oxidation was made in three different conditions, selected on the basis of previous preliminary evaluations [37,38]:

- In 0.5 M H_2SO_4 by applying 150 V for 2 min (label: ASD, anodic spark deposition);
- In NaF + Na_2SO_4 at 20 V, maintained constant for 6 h, followed by annealing at 400 °C for 2 h (label: A-NT, aqueous nanotubes);
- In NH_4F + ethylene glycol (EG) at 45 V maintained constant for 30 min, followed by annealing at 400 °C for 2 h (label: O-NT, organic nanotubes).

The effect of electrolyte has been taken into particular consideration, as it was found to have an important impact on TiO_2 nanostructuring and photocatalytic performance [38]. In order to induce the crystallization of nanotubular oxides, which are natively amorphous, annealing was performed at a temperature of 400 °C for 2 h. On the other hand, anodic oxides grown in ASD conditions already show a crystalline nature, therefore annealing was not performed, which allowed identify this preparation procedure as the least onerous from the point of view of both time and energy requirement.

2.2. Characterization

Scanning electron microscopy (SEM) was performed on a Carl Zeiss AG-EVO® Series 50 (Carl Zeiss AG, Oberkochen, Germany). Glow discharge optical emission spectroscopy (GDOES) was also performed on ASD samples to estimate oxide thickness, using a Jobin-Yvon RF-GDOES profilometer(Horiba, Fukuoka, Japan) Raman spectra were acquired using a InVia micro Raman spectrophotometer(Renishaw, Gloucestershire, UK) with 514.5 nm laser excitation wavelength and power on sample of about 1 mW.

2.3. RhB Photocatalytic Degradation

Tests were performed as reported in a previous work, by immersing a 6.0 ± 0.5 cm^2 sample in a beaker containing 25 mL of 10^{-5} M RhB [38] and irradiating samples for 6 h with a combined incandescent/Hg vapour lamp (Osram Vitalux, UV-A intensity of 3000 μW/cm^2, Munich, Germany). Dye discoloration was taken as reference of its degradation at its peak light absorption (550 nm), and was followed on a Spectronic 200E spectrophotometer. (Thermo Scientific, courtaboeuf, France) To ensure the dye molecule actually underwent breakup of the conjugated structure and not only deethylation [39], the whole wavelength range of deethylated products absorption (500–550 nm) was monitored: no deethylation was observed in none of the cases under investigation, therefore only data related to the 550 nm peak are presented.

RhB degradation was evaluated by measuring the dye concentration (C) over time using optical absorbance (*Abs*) and the Beer Lambert equation (Equation (1)):

$$Abs = \varepsilon \, l \, C \tag{1}$$

where l and ε are optical constants related to the measurement equipment (length of optical path) and reactant (optical absorption constant), respectively. Adsorption measurements performed in dark storage, monitoring in time the solution concentration after immersing the samples for a duration of 3 h, showed negligible adsorption of the dye onto the TiO$_2$ surface, as attested by absorbance values decreasing of approximately 1%. In photocatalytic tests, irradiation started after 20 min from sample immersion. Then data were processed considering a pseudo first order reaction kinetics, as typical of dye degradation reactions [38]. Data were then expressed as reaction rate constant per unit area (k_{app}) obtained from the following relationship (Equation (2)):

$$\frac{\ln\left(\frac{C}{C_0}\right)}{A} = -k_{app}t \tag{2}$$

where C_0 is initial RhB concentration, A is the sample surface area and t is irradiation time, in order to compare photocatalytic activity values normalized by each specimen area.

2.4. Toluene Photocatalytic Degradation

Toluene degradation (0.75 μmol m^{-3}) in air (25 °C and 50% RH) was assessed in a continuous stirred tank photoreactor (CSTR) operating at predetermined pollutant concentration. The desired toluene concentration inside the photochemical reactor was reached for each measurement by a two-step successive approximation process under UV irradiation, as detailed in a previous work [40]. This approach allows performing all the measurements at the same nominal operating concentration without any dependence on the actual sample activity. The computer-controlled experimental system includes an artificial air generator, an irradiation chamber that provides for photoreactor thermal and irradiance control and a GC/PID analyzer. The artificial air generator comprises four digital mass flow controllers with $\pm$ 0.25% nominal repeatability (model 5850, Brooks, Seattle, WA, USA) that deliver the required amount of pure nitrogen and oxygen from GC quality distribution lines (SIAD, Bergamo, Italy) and toluene (prediluted in nitrogen) from a standard cylinder (SIAD). Humidification is provided by a dedicated nitrogen line bubbled in bidistilled water. The reactor was operated with 1800 rpm mixing fan speed for all photocatalytic activity measurements. The irradiation chamber was equipped with two or six fluorescent lamps (PL-S/BLB, Philips, Amsterdam, The Netherlands) as UV-A source resulting respectively in 160 or 525 μW cm^{-2} irradiance at the sample surface (340–400 nm range), selected accordingly to the experimental needs. Toluene analyses were performed by an automated

GC/PID system provided with thermal desorption sampler/concentrator (GC955, Synspec, NL, Groningen, Netherlands). All gas samples were taken downstream the photoreactor after equilibration (i.e., in steady-state conditions); the reactor inlet concentration was measured at the reactor outlet with irradiation turned off. The photocatalytic activity was expressed as toluene area-specific degradation rate in air, as expressed in Equation (3) (r_a, in mol m^{-2} s^{-1}):

$$r_a = \left(\frac{Q}{A}\right)(C_0 - C) \tag{3}$$

where Q is the volumetric flow rate (m^3 s^{-1}); A is the sample area (m^2); C_0 is the inlet concentration of toluene (mol m^{-3}) and C its concentration in the reactor (mol m^{-3}).

The degradation rate measurement repeatability was calculated as error propagation in Equation (3) assuming $\pm$ 2% error in the concentration measurement C_0 and C, $\pm$ 1% error in the supply air volumetric flow Q and $\pm$ 3% error in the sample surface A [22], considering at least three tests per anodizing type.

3. Results

3.1. Anodic Oxides Characterization

SEM and Raman analyses were carried out to understand the anodic oxides morphology and crystal structure. Figure 1 (left) shows a microscopically homogeneous surface, characterized at higher magnification by a glassy appearance, big pores and cracks, characteristic of ASD samples [35,41]. Oxide thickness is approximately 0.7 μm, as attested by GDOES analysis. For A-NT with 6 h of anodizing time (Figure 1 center), vertical nanotubes stem from the substrate with about 100 nm of diameter. Non-uniform tube length is observed, as proved by the clear three-dimensional roughness observed in the higher magnification image. The cross-sectional views show that oxide nanotube thickness (tube length) is in the range of 1.5 μm, with thinner sections in some points, which is in the expected range of length for this kind of electrolyte, as already reported by other authors [42]. In organic solution (O-NT, Figure 1 right), from the top view image, no clear separation or detachment of single nanotubes is visible, apparently a nanoporous template rather than a nanotubular array was created, with non-uniform tube diameter ranging from tens to hundreds of nm. Yet, the cross section demonstrates that it actually consists of well-defined single tubes with smooth walls, where only the top part is continuous. Tube length is in the order of 3 μm and the thickness of the walls is about 35 nm (Table 1). For this reason, and for the sake of simplicity, these oxides will also be referred to as nanotubular. More details on the analysis of pore dimensions and distribution is reported in the discussion section.

Table 1. Summary of oxides characteristics as emerges from SEM images.

Sample	Electrolyte	Anodizing Time (min)	Avg Tube Wall Thickness (nm)	Avg Tube Length/Oxide Thickness (μm)
ASD	H_2SO_4	2	Does not apply	0.7
A-NT	$NaF + Na_2SO_4$	360	15	1.5
O-NT	$NH_4F + EG$	30	30–40	3

Structural characterization of the three kinds of anodic oxides was performed by Raman spectroscopy (Figure 2). ASD samples were analyzed without further treatment, confirming the predominance of anatase phase with sharp peaks at the expected positions (144, 399, 517, 639 cm^{-1}). Moreover, the high signal to noise ratio and the reduced width of the peaks attest a good degree of crystallinity of this anodic oxide. We underline also the presence of a further feature at about 430 cm^{-1}, which is compatible with the presence of rutile traces. Moving to nanotubes, both as prepared A-NT and O-NT present a band-like spectrum in the 100–800 cm^{-1} range (Figure 2a) typical of amorphous oxides, while after annealing at 400 °C the anatase peaks appear.

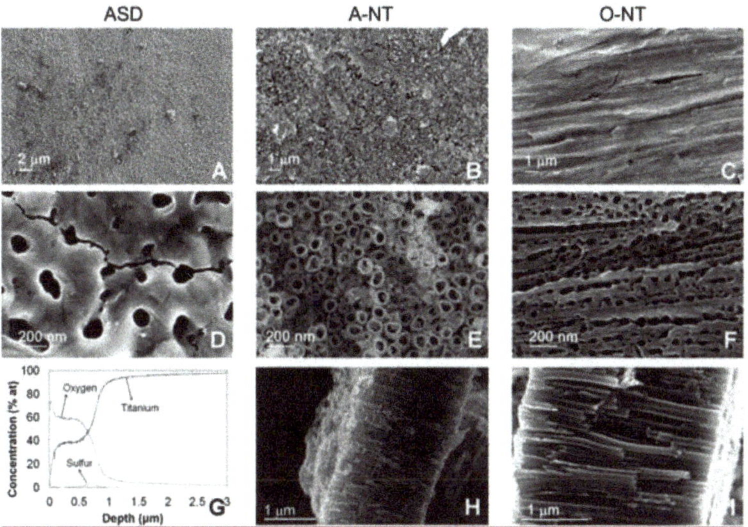

Figure 1. SEM micrographs of morphologies deriving from ASD (**left**), anodizing in aqueous solution with 4.5 h of anodizing (nanotubes, **center**) and in organic solution with 0.5 h of anodizing (nanopores/nanotubes, **right**) at different magnifications. The lower line reports information to derive oxide thickness: GDOES analysis for ASD samples, cross sections for nanotubes.

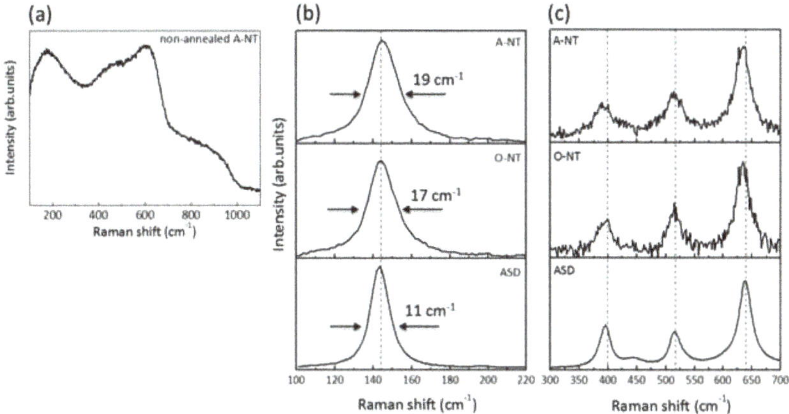

Figure 2. (**a**) Raman spectrum of non-annealed A-NT; (**b,c**) Raman spectra of the three samples in Table 1, as prepared ASD, A-NT and O-NT after annealing at 400 °C. Because of the very high intensity of the peak at 144 cm^{-1} with respect to all other peaks, spectra are separated into two panels with arbitrary intensity units. Vertical lines indicate reference peak positions for anatase crystal (144, 399, 517, 638 cm^{-1}). The width of the peak at 144 cm^{-1} is extracted by fitting with a lorentzian curve.

Comparing with ASD samples, the spectra of annealed nanotubes present lower signal to noise ratio and larger peaks, suggesting a lower degree of crystallinity. In particular, the width of the most intense peak (at 144 cm^{-1}), as reported in Figure 2, varies from 11 cm^{-1} for ASD samples to 17–19 cm^{-1} for nanotubes, confirming the presence of a larger amount of disorder [43]. It has been documented that TiO$_2$ crystalline structure has strong influence in the photocatalytic activity of the semiconductor, being anatase the most active phase [26]. On the other hand, the co-presence of anatase and a smaller fraction of rutile have also been proved beneficial for the overall TiO$_2$ photoactivity [44–46]. X-ray diffraction measurements were also performed to double check the crystalline structures present in the oxides, which indeed was confirmed as provided by Raman characterization (data not shown).

3.2. Photocatalytic Activity

Aqueous nanotubes (A-NT) were produced in a previous work, and showed good activity and reproducibility in RhB degradation [38]; therefore, these nanostructures were first considered. Since these nanotubes required a long anodizing (6 h), which makes the process particularly time and energy consuming, an attempt was made in the direction of reducing this parameter. For this reason, nanotubes produced in a shorter time (30 min) in organic solution and porous oxides obtained in a very quick ASD process lasting only 2 min are here proposed as alternatives. ASD was also previously evaluated as anodizing technique and gave interesting results concerning morphology and crystal structure [47]. A high voltage giving the co-presence of anatase (main phase) and rutile (to a minor extent) was chosen, as in previous literature this combination of crystal phases was observed to give better results than the sole anatase phase [48]. On the other hand, nanotubes in organic solution were shown to allow higher tube length compared to aqueous ones in shorter times [16]. Yet, the electrolyte poor conductivity generates high ohmic dissipations and consequent electrolyte heating during anodizing, which can lead to issues in reaching high voltages: therefore, a low cell voltage was chosen, i.e., 45 V, in order to have a more repeatable process and allow the anodizing of larger areas, i.e., in the order of tens of square centimeters. This is considered fundamental in the optimization of a surface treatment if already aiming at real applications, where the production of photocatalytic reactors for water or air purification are addressed, where nominal areas of photoactive items should overcome the few mm^2 that are often encountered in literature works.

Figure 3 shows the degradation kinetics of RhB over 6 h irradiation, either in presence of anodized samples or without any photocatalyst (reference test). In the reference test there is no dye degradation, only a small decrease in absorbance due to photolysis. TiO$_2$ samples all show a linear trend in ln(C/C$_0$), confirming the hypothesized pseudo-first order kinetics of RhB degradation. Concerning a comparison among different oxides, ASD and aqueous nanotubes produce similar photocatalytic activity, while organic nanotubes clearly show an enhanced efficiency in RhB degradation.

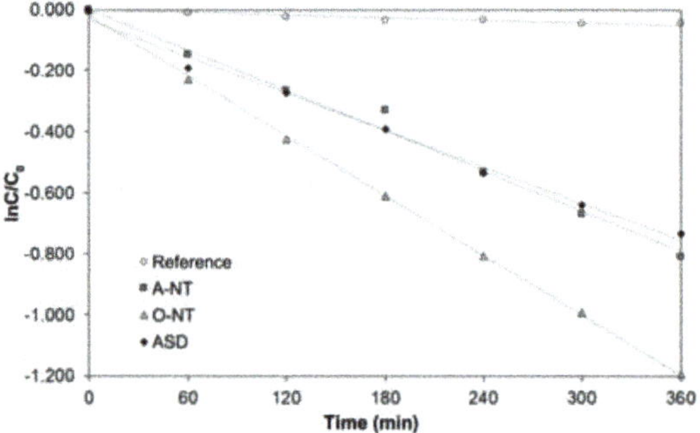

Figure 3. Kinetic plot for RhB degradation.

Table 2 reports the photocatalytic degradation rates of both RhB and toluene. A first consideration on the photocatalysis tests reproducibility is here provided. Rhodamine b-based photocatalysis measurements were all conducted at least in triple, and uncertainties on the resulting photocatalytic activity are of the same order of magnitude, for all samples, i.e., 10% of the measured activity. The toluene measurement system had been previously characterized, and instrumentation-related measurement errors are in the order of 8% on the actual measurement, as reported in Table 2 data uncertainty. Moreover, concerning reproducibility, series of four nanotubular specimens prepared in the same batch were measured, giving a relative standard deviation of the measured gas-phase photocatalytic activity that was non separable from the analytical method repeatability (~8% in the given conditions). This indicates a less than 8% sample repeatability inside a single preparative batch.

Table 2. Photocatalytic activity in gas and liquid phase for samples produced in different conditions: light intensity for toluene degradation 160 μW cm^{-2}, for RhB degradation 3000 μW cm^{-2}.

Activity Indicator	Reference	ASD	A-NT	O-NT
k_{app}, RhB (m^{-2} s^{-1})	0.002 ± 0.001	0.054 ± 0.005	0.050 ± 0.005	0.090 ± 0.010
r_a, toluene (nmol m^{-2} s^{-1})	0.00 ± 0.06	0.03 ± 0.06 [1]	0.14 ± 0.06	1.37 ± 0.11

[1] This measure was carried out at higher UV-A irradiance (525 μW cm^{-2}) in order to confirm the absence of any detectable activity.

A huge influence of anodizing electrolyte, and therefore oxide morphology, was observed especially in gas phase, where ASD samples show negligible photocatalytic activity in spite of an efficiency comparable to aqueous nanotubes in liquid phase. To the best of our knowledge, this observation has not been reported before. The implications of this behavior can suggest that a significant part of the photodegradation process lies in a balance between the effects of morphology of the nanostructure, most noticeably specific surface area and its accessibility, and presence, type and amount of crystalline structures. The variation of sample relative reactivity is also evident comparing the A-NT and O-NT samples, with a remarkably higher difference in gas phase (about a factor ten), while in liquid phase only a 40% difference was observed between the two nanostructures.

In order to give a rough comparison with a well-know photocatalyst, Figure 4 reports the gas-phase toluene degradation activities of screen-printed samples made with Aeroxide-P25 (Degussa) characterized in a previously described research [22]. Tests were repeated in conditions coherent with the present work. Data indicate that the toluene catalytic activity demonstrated by the O-NT sample correlates to the activity of a P-25 screen-printed sample with about 0.2~0.3 mg cm^{-2} TiO$_2$ layer loading. It is important to remark that such comparison is merely indicative, as it depends on the deposition technique conditions and consequent TiO$_2$ layer parameters, such as degree of compaction and specific surface area available, which in turn influence the interaction with the effluent. Therefore this comparison holds only for the given measurement conditions.

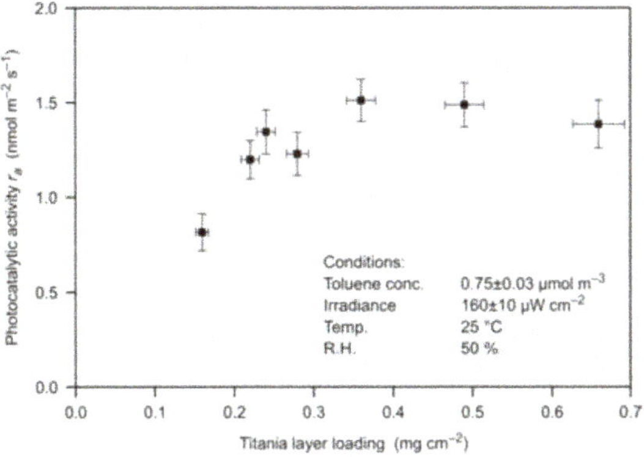

Figure 4. Photocatalytic toluene degradation activity of screen-printed titania samples made with Degussa Aeroxide P25 [22] in the experimental conditions indicated inside the figure.

4. Discussion

The presented results emphasize the variations in relative activity demonstrated by similar morphologies (i.e., nanotubes or pores) when utilized in gas or liquid phase photocatalytic processes. The different behaviors are possibly due to dissimilar physical constraints of the oxide matrices (e.g., reactant diffusivity), and they appear more conspicuous in the gas system, indicating a different exploitation of the catalyst in gas and liquid phase reactions.

It is worth comparing first the photocatalytic activity of the three materials in liquid phase. In this environment, differences are flattened: the overall dye degradation extent after 6 h of test ranges from 50% (ASD, A-NT) to 65% (O-NT), giving a difference of only 25% between the most and least effective oxides on the overall degradation percent, which increases to 40% if the reaction constants are taken into account. As previously pointed out, nanotubes produced in aqueous electrolyte and porous oxides obtained via anodic spark deposition present very similar activities, in spite of differences both in the crystallinity and in the surface area of the oxides. Yet, ASD samples exhibit higher crystallinity, which is beneficial to improve the separation and transport of photogenerated charge carriers, and also present a mixture of anatase and rutile phases which, as abovementioned, may show benefits compared to the pure anatase form [44–46]. On the other hand, nanotubes present a larger surface area, with double oxide thickness and larger amount of pores, in spite of a partial obstruction of some nanotubes. This is more clearly visible in Figure 5, where the surface topography has been re-addressed evidencing the darker areas—pore or tube openings—that contribute to the increase in surface area of these nanostructures with respect to the nominal one. The image was analyzed with Gwyddion [49] and pore size distribution was extracted and reported in Figure 6. Data analysis revealed a percentage of surface covered by pores triple in the A-NT sample, i.e., nanotubes produced in aqueous electrolyte, compared to ASD-obtained sample (32% vs. 10%, respectively), with an intermediate condition for the O-NT oxide grown in the organic electrolyte (19%).

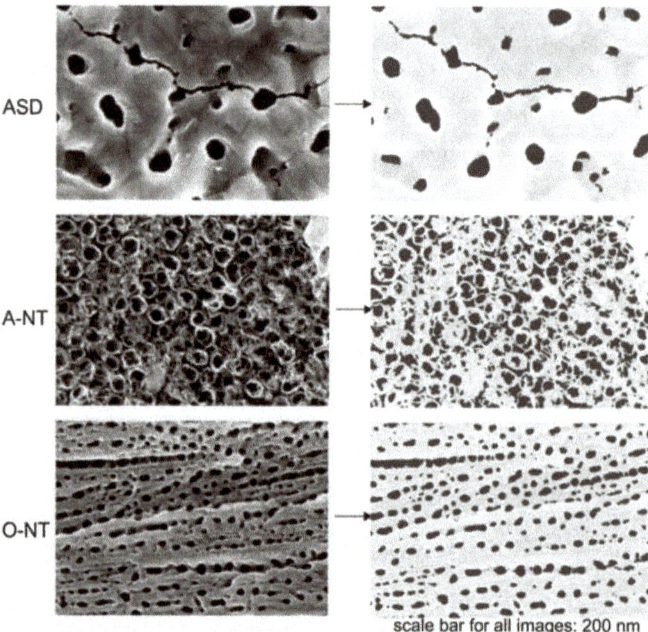

Figure 5. Topography of samples and highlight of open porosities.

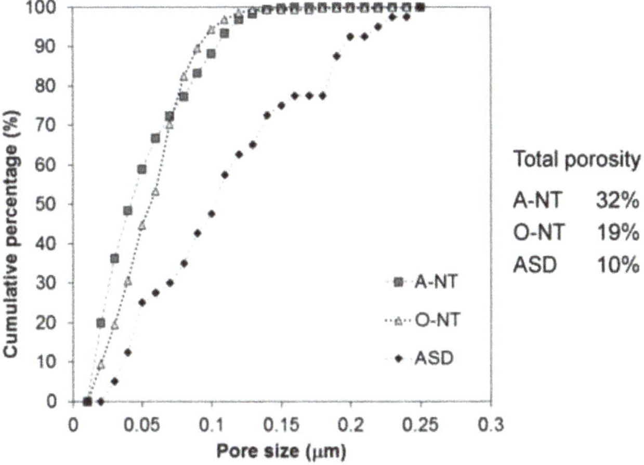

Figure 6. Analysis of porosities appearing on the three oxides surfaces.

Therefore, the A-NT oxide can count on an overall double thickness and higher percent of pores with respect to the total oxide surface with respect to ASD: this increase in specific surface area is sufficient to counterbalance the lower oxide crystallinity.

On the other hand, O-NT samples also have a smaller percent of pore openings with respect to A-NT and same crystallinity (Figure 2), and should therefore present a lower photocatalytic activity, while on the opposite they showed the best performances in liquid phase photocatalysis. This was explained by considering the cross section of O-NTs, which is clearer from obstacles with smoother walls, as explained in the following.

In fact, Figure 7 reports a schematic representation of nanotubes cross-section highlighting the tortuosity of A-NT, which may be responsible of a shorter diffusion path of RhB molecules in aqueous nanotubes. It is evident that, in spite of a smaller overall number of pores, O-NT pores are more regular and freely accessible, with reference to the frequent partial pore occlusion visible in A-NT in Figure 4, which causes the appearance of a large percent of small porosity and the consequent reduction in average pore size. The difficulty for RhB molecules to diffuse through small pores and until the bottom of nanotubes would account for the higher photoactivity of nanotubes obtained in organic solution. Finally, ripples formation on the wall of nanotubes not only may obstruct available pores, but also act as charge carrier recombination points, therefore reducing activity: also this feature is more pronounced in nanotubes produced in aqueous electrolyte (A-NT) [50].

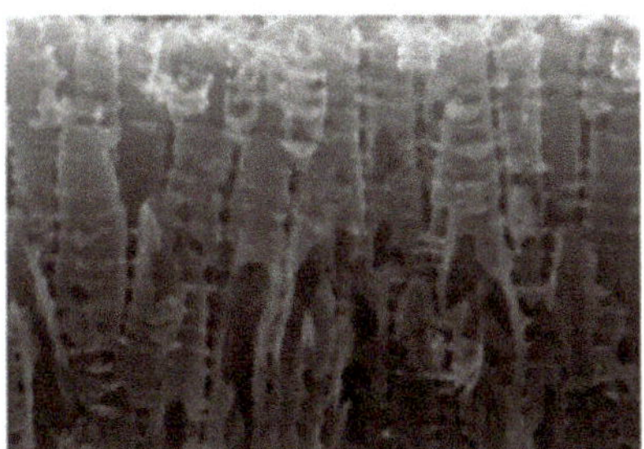
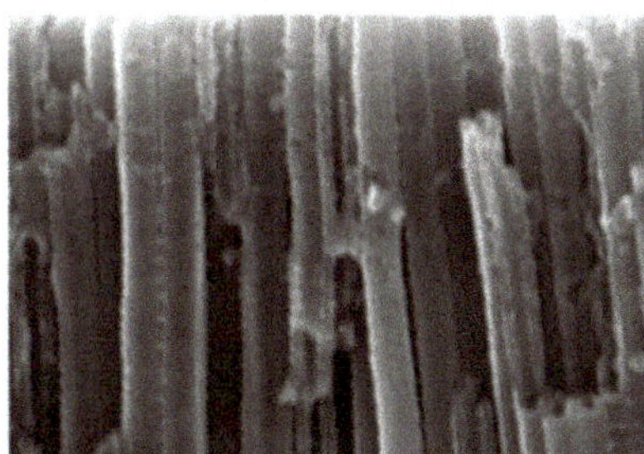

Figure 7. Schematic representation of nanotubes cross-section highlighting the tortuosity of A-NT.

In gas phase, different considerations hold. In fact, toluene molecules are way smaller and have easier mobility compared with RhB ones. In this way, nanotubes depth is exploited to a larger extent: it is not surprising then to see much higher photoactivity for O-NT, where tubes are longer, smoother, less obstructed and present less recombination centers compared with A-NT. On the other hand, samples prepared by ASD show negligible activity in this case, since they are strongly penalized by the extremely low surface area.

Such observations underline the utility of having a set of photocatalysts with different morphological and structural characteristics available, since different reactants may find easier decomposition over given substrates but not on others, or differences could be attenuated depending on the reactant nature and reaction paths. This could allow the use of more easy-to-produce ASD photocatalysts in light duty occurrence, such as RhB degradation, while more active photocatalysts, such as organic nanotubes, would be required in more demanding conditions such as gas phase pollutants degradation.

5. Conclusions

We presented the production and characterization of nanostructured titanium dioxide photocatalysts, obtained by means of different anodic oxidation techniques. Nanotubes prepared in an organic electrolyte show good photoactivity, in both gas and liquid phase.

The operating parameters affect the morphology and surface area of the different TiO_2 nanostructures, and thus their efficiency in pollutants degradation. The effect of the preparation electrolyte on photocatalytic activity is evident: in particular, films with nanotubular morphology present much higher efficiency if produced in organic electrolytes than in aqueous ones, especially in gas phase, where the pollutant can better diffuse through the longer length of the organic nanotubes. The morphology obtained by anodizing at high voltage in absence of fluorides presents a comparable efficiency with aqueous nanotubes when used in liquid phase, and no activity when used in gas phase, in spite of a superior crystallinity: this suggests that surface area is the ruling parameter in gas phase photocatalysis, while in liquid phase, where larger molecules are generally involved—e.g., organic dyes—only a part of the pores length can be exploited, therefore flattening the differences in photocatalytic activities among the three oxides.

Author Contributions: M.P.P., M.V.D. and B.D.C. conceived and designed the experiments; B.E.S. prepared the samples and performed the rhodamine B photocatalysis experiments; A.L.B. and V.R. performed SEM and Raman measurements and analyzed Raman data; A.S. and L.S. performed the toluene photocatalysis experiments; M.V.D. analyzed SEM data; M.V.D. and B.E.S. wrote the paper.

Conflicts of Interest: The authors declare no conflict of interest.

References

1. Ibhadon, A.; Fitzpatrick, P. Heterogeneous photocatalysis: Recent advances and applications. *Catalysts* **2013**, *3*, 189–218. [CrossRef]
2. Lazar, M.; Varghese, S.; Nair, S. Photocatalytic water treatment by titanium dioxide: Recent updates. *Catalysts* **2012**, *2*, 572–601. [CrossRef]
3. Tompkins, D.; Lawnicki, B.; Zeltner, W.; Anderson, M. Evaluation of photocatalysis for gas-phase air cleaning—Part 1: Process, Technical, and Sizing Considerations. *ASHRAE* **2005**, *111*, 60–84.
4. Hsien, H.-Y.; Chang, C.-F.; Chen, Y.-H.; Cheng, S. Photodegradation of aromatic pollutants in water over TiO_2 supported on molecular sieves. *Appl. Catal. B Environ.* **2001**, *31*, 241–249. [CrossRef]
5. Einaga, H.; Mochiduki, K.; Teraok, Y. Photocatalytic oxidation processes for toluene oxidation over TiO_2 catalysts. *Catalysts* **2013**, *3*, 219–231. [CrossRef]
6. Piera, E.; Ayllón, J.; Doménech, X.; Peral, J. TiO_2 deactivation during gas-phase photocatalytic oxidation of ethanol. *Catal. Today* **2002**, *76*, 259–270. [CrossRef]
7. Lisebigler, A.; Lu, G.; Yates, J. Photocatalysis on TiO_2 surfaces: Principles, mechanisms and selected results. *Chem. Rev.* **1995**, *95*, 735–758. [CrossRef]

8. Gunti, S.; Kumar, A.; Ram, M.K. Nanostructured photocatalysis in the visible spectrum for the decontamination of air and water. *Int. Mater. Rev.* **2018**, *63*, 257–282. [CrossRef]

9. Khalid, N.R.; Majid, A.; Tahir, M.B.; Niaz, N.A.; Khalid, S. Carbonaceous-TiO$_2$ nanomaterials for photocatalytic degradation of pollutants: A review. *Ceram. Int.* **2018**, *43*, 14552–14571. [CrossRef]

10. Han, E.; Vijayarangamuthu, K.; Youn, J.-S.; Park, Y.-K.; Jung, S.-C.; Jeon, K.-J. Degussa P25 TiO$_2$ modified with H$_2$O$_2$ under microwave treatment to enhance photocatalytic properties. *Catal. Today* **2018**, *303*, 305–312. [CrossRef]

11. Melcher, J.; Feroz, S.; Bahnemann, D. Comparing photocatalytic activities of commercially available iron-doped and iron-undoped aeroxide TiO$_2$ P25 powders. *J. Mater. Sci.* **2018**, *52*, 6341–6348. [CrossRef]

12. Souza, R.P.; Freitas, T.K.F.S.; Domingues, F.S.; Pezoti, O.; Ambrosio, E.; Ferrari-Lima, A.M.; Garcia, J.C. Photocatalytic activity of TiO$_2$, ZnO and Nb$_2$O$_5$ applied to degradation of textile wastewater. *J. Photochem. Photobiol. A Chem.* **2016**, *329*, 9–17. [CrossRef]

13. Alvarez-Corena, J.R.; Bergendahl, J.A.; Hart, F.L. Advanced oxidation of five contaminants in water by UV/TiO$_2$: Reaction kinetics and byproducts identification. *J. Environ. Manag.* **2016**, *181*, 544–551. [CrossRef] [PubMed]

14. Qin, X.; Jing, L.; Tian, G.; Qu, Y.; Feng, Y. Enhanced photocatalytic activity for degrading Rhodamine B solution of commercial Degussa P25 TiO$_2$ and its mechanisms. *J. Hazard. Mater.* **2009**, *172*, 1168–1174. [CrossRef] [PubMed]

15. Ahmed, S.; Rasul, M.; Martens, W.; Brown, R.; Hashib, M. Advances in heterogeneous photocatalytic degradation of phenols and dyes in wastewater: A review. *Water Air Soil Pollut.* **2011**, *215*, 3–29. [CrossRef]

16. Paramasivam, I.; Jha, H.; Liu, N.; Schmuki, P. A Review of photocatalysis using Self-organized TiO$_2$ Nanotubes and Other Ordered Oxide Nanostructures. *Small* **2012**, *8*, 3073–3103. [CrossRef] [PubMed]

17. Grzechulska-Damszel, J.; Mozia, S.; Morawski, A. Integration of photocatalysis with membrane processes for purification of water contaminated with organic dyes. *Catal. Today* **2010**, *156*, 295–300. [CrossRef]

18. Sarantopoulos, C.; Puzenat, E.; Guillard, C.; Herrmann, J.-M.; Gleizes, A.; Maury, F. Microfibrous TiO$_2$ supported photocatalysts prepared by metal-organic chemical vapor infiltration for indoor air and waste water purification. *Appl. Catal. B Environ.* **2009**, *91*, 225–233. [CrossRef]

19. Kathaee, A.; Kasiri, M. Photocatalytic degradation of organic dyes in the presence of nanostructured titanium dioxide: Influence of the chemical structure of dyes. *J. Mol. Catal. A Chem.* **2010**, *328*, 8–26. [CrossRef]

20. Bettini, L.G.; Diamanti, M.V.; Sansotera, M.; Pedeferri, M.P.; Navarrini, W.; Milani, P. Immobilized TiO$_2$ nanoparticles produced by flame spray for photocatalytic water remediation. *J. Nanoparticle Res.* **2016**, *18*, 238. [CrossRef]

21. Persico, F.; Sansotera, M.; Bianchi, C.L.; Cavallotti, C.; Navarrini, W. Photocatalytic activity of TiO$_2$-embedded fluorinated transparent coating for oxidation of hydrosoluble pollutants in turbid suspensions. *Appl. Catal. B Environ.* **2015**, *170–171*, 83–89. [CrossRef]

22. Strini, A.; Sanson, A.; Mercadelli, E.; Sangiorgi, A.; Schiavi, L. Low irradiance photocatalytic degradation of toluene in air by screen-printed titanium dioxide layers. *Thin Solid Films* **2013**, *545*, 537–542. [CrossRef]

23. Gnatyuk, Y.; Smirnova, N.; Korduban, O.; Eremenko, A. Effect of zirconium incorporation on the stabilization of TiO$_2$ mesoporous structure. *Surf. Interface Anal.* **2010**, *42*, 1276–1280. [CrossRef]

24. Strini, A.; Sanson, A.; Mercadelli, E.; Bendoni, R.; Marelli, M.; Dal Santo, V.; Schiavi, L. In-situ anatase phase stabilization of titania photocatalyst by sintering in presence of Zr^{4+} organic salts. *Appl. Surf. Sci.* **2015**, *347*, 883–890. [CrossRef]

25. Diamanti, M.V.; Ormellese, M.; Pedeferri, M. Application-wise nanostructuring of anodic films on titanium: A review. *J. Exp. Nanosci.* **2015**, *10*, 1285–1308. [CrossRef]

26. Macak, J.; Zlamal, M.; Krysa, J.; Schmuki, P. Self-organized TiO$_2$ nanotube layer as highly efficient photocatalysis. *Small* **2007**, *3*, 300–304. [CrossRef] [PubMed]

27. Riboni, F.; Nguyen, N.T.; So, S.; Schmuki, P. Aligned metal oxide nanotube arrays: Key-aspects of anodic TiO$_2$ nanotube formation and properties. *Nanoscale Horiz.* **2016**, *1*, 445–466. [CrossRef]

28. Lee, K.; Mazare, A.; Schmuki, P. One-dimensional titanium dioxide nanomaterials: Nanotubes. *Chem. Rev.* **2014**, *114*, 9385–9454. [CrossRef] [PubMed]

29. Qin, L.; Chen, Q.; Lan, R.; Jiang, R.; Quan, X.; Xu, B.; Zhang, F.; Jia, Y. Effect of anodization parameters on morphology and photocatalysis properties of TiO$_2$ Nanotube Arrays. *J. Mater. Sci. Technol.* **2015**, *31*, 1059–1064. [CrossRef]

30. Xie, Z.; Blackwood, D. Effects of anodization parameters on the formation of titania nanotubes in ethylene glycol. *Electrochim. Acta* **2010**, *56*, 905–912. [CrossRef]

31. Raja, K.; Gandhi, T.; Misra, M. Effect of water content of ethylene glycol as electrolyte for synthesis of ordered titania nanotubes. *Electrochem. Commun.* **2007**, *9*, 1069–1076. [CrossRef]

32. Haring, A.; Morris, A.; Hu, M. Controlling morphological parameters of anodized titania nanotubes for optimized solar energy applications. *Mater* **2012**, *5*, 1890–1909. [CrossRef]

33. Mena, E.; de Vidales, M.J.; Mesones, S.; Marugán, J. Influence of anodization mode on the morphology and photocatalytic activity of TiO_2-NTs array large size electrodes. *Catal. Today* **2018**, in press. [CrossRef]

34. Ye, Y.; Feng, Y.; Bruning, H.; Yntema, D.; Rijnaarts, H.H.M. Photocatalytic degradation of metoprolol by TiO_2 nanotube arrays and UV-LED: Effects of catalyst properties, operational parameters, commonly present water constituents, and photo-induced reactive species. *Appl. Catal. B Environ.* **2018**, *220*, 171–181. [CrossRef]

35. Diamanti, M.V.; Ormellese, M.; Pedeferri, M. Alternating current anodizing of titanium in halogen acids combined with anodic spark depositions: Morphological and structural variations. *Corros. Sci.* **2010**, *52*, 1824–1829. [CrossRef]

36. Zhuiykov, S. *Nanostructured Semiconductor Oxides for the Next Generation of Electronics and Functional Devices. Properties and Applications, 1st ed*; Woodhead Publishing: Cambridge, UK, 2013.

37. Xiao, P.; Zhang, Y.; Garcia, B.S.S.; Liu, D.C.G. Nanostructured electrode with titania nanotubes arrays: Fabrication, electrochemical properties, and applications for biosensing. *J. Nanosci. Nanotechnol.* **2008**, *8*, 1–11. [CrossRef]

38. Diamanti, M.V.; Ormellese, M.; Marin, E.; Lanzutti, A.; Mele, A.; Pedeferri, M. Anodic titanium oxide as immobilized photocatalyst in UV or visible light devices. *J. Hazard. Mater.* **2011**, *186*, 2103–2109. [CrossRef] [PubMed]

39. Wu, J.-M.; Zhang, T.-W. Photodegradation of rhodamine B in water assisted by titania films prepared through a novel procedure. *J. Photochem. Photobiol. A Chem.* **2004**, *162*, 171–177. [CrossRef]

40. Strini, A.; Schiavi, L. Low irradiance toluene degradation activity of a cementitious photocatalytic material measured at constant pollutant concentration by a successive approximation method. *Appl. Catal. B Environ.* **2011**, *103*, 226–231. [CrossRef]

41. Park, Y.-J.; Shin, K.-H.; Song, H.-J. Effects of anodizing conditions on bong strength of anodically oxidized film on titanium substrate. *Appl. Surf. Sci.* **2007**, *253*, 6013–6018. [CrossRef]

42. Macak, J.; Sirotna, K.; Schmuki, P. Self-organized porous titanium oxide prepared in Na_2SO_4/NaF electrolytes. *Electrochim. Acta* **2005**, *50*, 3679–3684. [CrossRef]

43. Li Bassi, A.; Cattaneo, D.; Russo, V.; Bottani, C.E.; Barborini, E.; Mazza, T.; Piseri, P.; Milani, P.; Ernst, F.O.; Wegner, K.; et al. Raman Spectroscopy Characterization of titania nanoparticles produced by flame pyrolysis: The influence of size and stoichiometry. *J. Appl. Phys.* **2005**, *98*, 074305. [CrossRef]

44. Ohno, T.; Sarukawa, K.; Tokieda, K.; Matsumura, M. Morphology of a TiO_2 photocatalyst (Degussa, P-25) consisting of anatase and rutile crystalline phases. *J. Catal.* **2001**, *203*, 82–86. [CrossRef]

45. Siah, W.R.; Lintang, H.O.; Shamsuddin, M.; Yuliati, L. High photocatalytic activity of mixed anatase-rutile phases on commercial TiO_2 nanoparticles. *IOP Conf. Ser. Mater. Sci. Eng.* **2016**, *107*, 012005. [CrossRef]

46. Hurum, D.C.; Agrios, A.G.; Gray, K.A.; Rajh, T.; Thurnauer, M.C. Explaining the enhanced photocatalytic activity of Degussa P25 mixed-phase TiO_2 using EPR. *J. Phys. Chem. B* **2003**, *107*, 4545–4549. [CrossRef]

47. Diamanti, M.V.; Pedeferri, M. Effect of anodic oxidation parameters on the titanium oxides formation. *Corros. Sci.* **2007**, *49*, 939–948. [CrossRef]

48. Bakardjieva, S.; Šubrt, J.; Štengl, V.; Dianez, M.J.; Sayagues, M.J. Photoactivity of anatase-rutile TiO_2 nanocrystalline mixtures obtained by heat treatment of homogeneously precipitated anatase. *Appl. Catal. B Environ.* **2005**, *58*, 193–202. [CrossRef]

49. Nečas, D.; Klapetek, P. Gwyddion: An open-source software for SPM data analysis. *Cent. Eur. J. Phys.* **2012**, *10*, 181–188. [CrossRef]

50. Lynch, R.P.; Ghicov, A.; Schmuki, P. A photo-electrochemical investigation of self-organized TiO_2 nanotubes. *J. Electrochem. Soc.* **2010**, *157*, G76. [CrossRef]

© 2018 by the authors. Licensee MDPI, Basel, Switzerland. This article is an open access article distributed under the terms and conditions of the Creative Commons Attribution (CC BY) license (http://creativecommons.org/licenses/by/4.0/).

 *materials*

Article

Sol-Gel Hydrothermal Synthesis and Visible Light Photocatalytic Degradation Performance of Fe/N Codoped TiO_2 Catalysts

Hsu-Hui Cheng [1,2,*] , Shiao-Shing Chen [3], Sih-Yin Yang [3], Hui-Ming Liu [1] and Kuang-Shan Lin [3]

[1] Department of Safety, Health and Environmental Engineering, Hungkuang University, Taichung 43302, Taiwan; hmliu@sunrise.hk.edu.tw
[2] Hsuteng Consulting International Co., Ltd. Taichung 40764, Taiwan
[3] Institute of Environment Engineering and Management, National Taipei University of Technology, Taipei 10643, Taiwan; f10919@ntut.edu.tw (S.-S.C.); lovey1124@gmail.com (S.-Y.Y.); chenghh1126@gmail.com (K.-S.L.)
* Correspondence: hhcheng1126@gmail.com; Tel.: +886-4-2461-6260

Received: 12 April 2018; Accepted: 30 May 2018; Published: 3 June 2018

Abstract: Using $Ti(OC_4H_9)_4$ as a precursor, $Fe(NO_3)_3 \cdot 9H_2O$ as the source of iron, and NH_4NO_3 as the source of nitrogen, an Fe/N codoped TiO_2 catalyst was prepared using a sol-gel hydrothermal method. The as-prepared powders were characterized using X-ray powder diffraction, electron spectroscopy for chemical analysis, Fourier-transform infrared spectroscopy, and ultraviolet-visible spectrophotometry. Fe and N codoping resulted in decreased crystallite size and increased specific surface area. Results of the photocatalytic degradation of acid orange 7 (AO7) in a continuous-flow fluidized-bed reactor indicated that the maximum decolorization (more than 90%) of AO7 occurred with the Fe/N-TiO_2 catalyst (dosage of 20 g/L) when a combination of visible light irradiation for 10 h HRT (hydraulic retention time), and a heterogeneous system was used. The AO7 degradation efficiency was considerably improved by increasing the hydraulic retention time from 2.5 to 10 h or by reducing the initial AO7 concentration from 300 to 100 mg/L. The reaction rate increased with the light intensity and the maximum value occurred at 35 mW/cm^2; moreover, the efficiency of the AO7 degradation increased when the pH decreased with maximum efficiency at pH 3.

Keywords: Fe/N-TiO_2; sol-gel; hydrothermal; photocatalytic; visible-light

---

1. Introduction

Environmental pollution is a considerable concern in the modern world. An estimated 2% of dyes produced annually are discharged as effluents from manufacturing plants, whereas 10% of dyes are discharged from textile and related industries [1]. Effluents generated from textile manufacturing contain a variety of pollutants characterized by deep coloration, high oxygen demand, high pH, large amounts of suspended solids, and low or nonbiodegradability [2,3]. Many methods have been tested to remove dyes from industrial effluents, including biological processes, adsorption, and coagulation. However, these methods still generate a large amount of sludge or solid waste that requires further treatment.

Advanced oxidation processes are a suitable alternative to traditional methods for solving environmental problems caused by the discharge of textile-dyeing wastewater. Titanium dioxide (TiO_2) is a heterogeneous photocatalysts and TiO_2 based photocatalysis is a promising technique for wastewater treatment [4], especially for wastewater containing refractory organic compounds. However, the large band gap for highly oriented TiO_2 powders with pure anatase structure and rutile are 3.2 and 3.0 eV, respectively. Therefore, pure TiO_2 can absorb solar light only in the near ultraviolet (UV) region. To modify this property and shift the excitation threshold toward higher wavenumbers, the recombination time of free radicals must be extended or the phase composition must be changed to significantly affect the optical and electrical properties of the material. Doping with different nonmetallic or metallic elements has often been employed to improve photocatalytic activity [5].

Among nonmetallic dopants, doping TiO_2 with N is one of the most effective methods to produce effects from visible light irradiation [6]. However, because the N 2p states are strongly localized at the top of the valence band, the photocatalytic efficiency of N-doped TiO_2 decreases. The isolated empty states tend to trap photogenerated electrons, thereby reducing the photogenerated current [7]. Doping TiO_2 with two different elements, namely nitrogen and cheaper Fe ions, has attracted interest in computational studies. Several papers have reported that Fe ions can trap holes or electrons at low doping levels, whereas they become recombination centers at high doping levels [8–10]. The photocatalytic activities of these powders are approximately two to four times higher than those of pure anatase TiO_2 under visible light irradiation. The synthesis of TiO_2 nanoparticles by using a combination of sol-gel and hydrothermal methods is another recent innovation. The sol-gel hydrothermal method combines the advantages of the sol-gel method with high-pressure hydrothermal conditions [11]; particle size and morphology can be controlled during the hydrothermal process [11,12].

In this paper, we present a sol-gel hydrothermal method for the fabrication of Fe/N-TiO_2 catalysts that respond to visible light. The photocatalytic activity of Fe/N-TiO_2 was measured for the degradation of acid orange 7 (AO7) in a continuous-flow fluidized-bed system under visible light irradiation. The effects of operational parameters, such as the catalyst activity, dosage, and solution pH, were also examined.

2. Materials and Methods

2.1. Sample Preparation

Fe/N-TiO_2 was prepared using a sol-gel hydrothermal method. A suitable amount (0.1 mol) of titanium tetra-n-butoxide $[Ti(OC_4H_9)_4]$ (Sigma Aldrich, MO, USA) was dissolved in 100 mL of anhydrous ethanol (Merck, Darmstadt, Germany) to obtain solution A. Moreover, 0.0012 mol of iron nitrate $[Fe(NO_3)_3 \cdot 9H_2O]$ (Merck, Darmstadt, Germany) and 0.001 mol of ammonium nitrate (Merck, Darmstadt, Germany) were mixed with 2 mL of distilled water and 10 mL of acetic acid (Merck, Darmstadt, Germany) to prepare solution B. Then, solution A was slowly added to solution B at a rate of 2 mL per minute under stirring for up to 48 h. The sample mixture was transferred to a hydrothermal flask to undergo treatment at 100, 150, 175, and 200 °C for 1 h. The resulting Fe/N-TiO_2 powder was washed with distilled water until a pH of 7 was established and then dried at 80 °C for 24 h.

2.2. Characterization

The band gap of Fe/N-TiO_2 was measured using a UV-visible spectrophotometer (Cary 300 Bio, Varian, Mulgrave, Victoria, Australia) equipped with an integrating sphere for diffuse reflectance spectra. The chemical composition of Fe/N-TiO_2 was verified through electron spectroscopy for chemical analysis (ESCA; ESCALAB 250, VG Scientific, UK). Crystal structures were obtained through X-ray diffraction (XRD; Rigaku Co. DMAX 2200VK, Tokyo, Japan) using Cu Kα radiation (λ = 1.5418 Å). All peaks measured through XRD were assigned by comparison with those of the Joint Committee on Powder Diffraction Standards (JCPDS 04-002-2678) [13]. The specific surface area (BET, m^2 g^{-1}) was calculated using the BET equation, and total pore volume (Vt, m^3 g^{-1}) was evaluated by converting the adsorption amount at P/P_0 = 0.95 to the volume of the liquid adsorbate.

2.3. Photocatalytic Experiments

The upflow fluidized-bed system is shown in Figure 1. The photocatalytic activities of Fe/N-TiO_2 samples under visible light were evaluated based on the degradation rate of AO7 in a cylindrical quartz reactor (40/30 mm OD/ID; height = 500 mm) containing 20 g of Fe/N-TiO$_2$ and 5 L of a 200 mg/L AO7 aqueous solution. The photoreactor was open to the atmosphere, and the quartz reactor was surrounded by 14 light tubes. The visible light tubes were germicidal lamps with a wavelength of 419 nm (Sankyo Denki, Tokyo, Japan). The light power (approximately 8 mW/cm^2) in the center of the reactor in air was measured using a hand-held optical power meter (Model 840-C, Newport, Irvine, CA, USA). The photodegradation rates of AO7 solutions were determined by periodically measuring the absorbance at λ = 484 nm by using a Hach DR 4000 UV-visible spectrophotometer (Hach, Loveland, CO, USA).

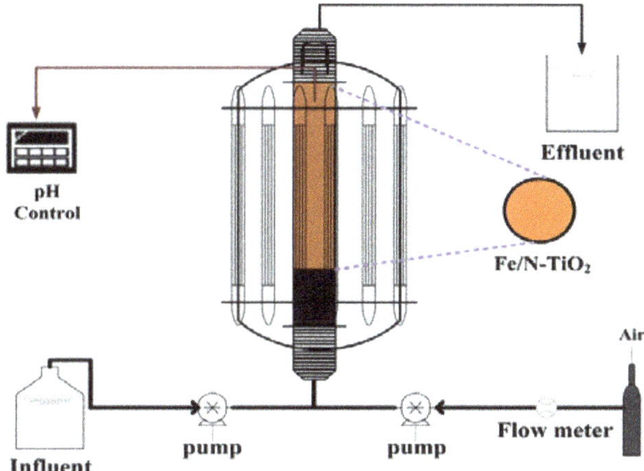

Figure 1. Schematic of the upflow fluidized-bed system.

3. Results and Discussion

3.1. Characterization of the N/Fe-TiO$_2$ Samples

Figure 2a presents the XRD patterns of undoped (TiO$_2$) and Fe/N-TiO$_2$ particles as a function of the reaction temperature. Fe/N-TiO$_2$ particles were readily indexed to the diffraction peaks of the anatase phase (JCPDS 04-002-2678) and exhibited the presence of an intense peak corresponding to the (101) plane. The major peaks observed corresponded to the (101), (004), (200), (105), and (204) planes of the anatase phase [14]. For 100, 150, 175, and 200 °C Fe/N-TiO$_2$ particles, the crystallite sizes were 10.65, 10.79, 12.11, and 13.46 nm, respectively. Smaller crystallite sizes were obtained for the codoped samples, which indicated that the incorporation of Fe and N ions restricted the growth of TiO$_2$ crystallite and prevented the transformation of anatase to rutile [15]. Deng et al. [16] also investigated the morphology of Fe-doped titania nanotubes synthesized using the sol-gel and hydrothermal methods. They found that the addition of Fe slowed the crystallization process and prevented the growth of crystallite TiO$_2$. The crystallite size of Fe/N-TiO$_2$ particles increased with the reaction temperature (Table 1).

To determine whether codoping with Fe/N was successful, the surface of Fe/N-TiO$_2$ composites was examined through ESCA. The ESCA spectra of Ti 2p in Fe/N-TiO$_2$ shown in Figure 2(b) reveal that the Ti 2p$_{1/2}$ and Ti 2p$_{3/2}$ peaks at 464.2 and 458.5 eV, respectively, were in a favorable agreement with those previously observed for Ti^{4+} [17].The presence of N in TiO$_2$ particles was substantiated by the N 1s spectra and significant peaks around 400 eV, which can be attributed to the formation of anionic N in O$-$Ti$-$N linkages [18], whereas the iron peak (710 eV) was attributed to Fe^{3+}, indicating the formation of Fe$_2$O$_3$ [19]. Saha and Tompkins [20] investigated N 1s ESCA spectra during the oxidation process of Ti–N and assigned the peaks at 400 eV to be molecularly chemisorbed $\gamma-N_2$. Kim et al. [15]

reported that the ionic radii of Fe^{3+} (0.64 Å) and Ti^{4+} (0.68 Å) are similar and that Fe^{3+} can therefore be incorporated into the lattice of TiO_2 to form a Ti–O–Fe bond in $Fe/N-TiO_2$. The results indicate that Fe is present in the form of Fe^{3+} by replacing Ti^{4+} in the doped photocatalyst, which may change the charge distribution of atoms on the photocatalyst surface, resulting in enhanced photocatalytic activity. By contrast, the decrease of Ti binding energy upon N-doping could be interpreted as the formation of O–Ti–N in the TiO_2 lattice [19], which indicates that nitrogen incorporation can successfully retard the charge recombination at the TiO_2/dye/electrolyte interface. Additionally, the concentrations of Fe and N in $Fe/N-TiO_2$ determined using ESCA were 5.58 and 5.48 wt %, respectively, which were consistent with the theoretical expectation.

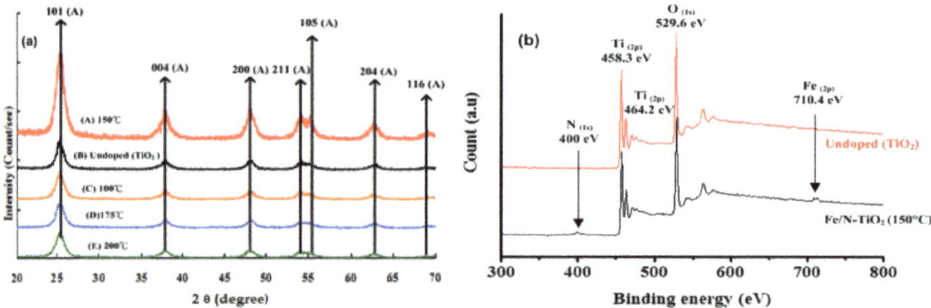

Figure 2. $Fe/N-TiO_2$ powders prepared using the sol-gel hydrothermal process: (**a**) XRD patterns of samples A–E, (**b**) ESCA spectra of the samples.

Table 1. Physicochemical properties of doped and undoped samples.

Sample		Crystallite Size (nm)	BET Surface Area (m^2 g^{-1})	Band Gap (eV)	Degradation (%) (Batch-Type)
Undoped TiO$_2$		30.01	56	3.20	31
Fe/N-TiO$_2$	100 °C	10.65	233	2.67	61
	150 °C	10.79	226	2.67	95
	175 °C	12.11	211	2.74	93
	200 °C	13.46	213	2.55	67

The calculation of the band gap of materials can be conducted using the following formulation: absorption coefficient *(a)* and the incident photon energy *(hv)* can be written as $a = Bi \cdot (hv - E_g)^2 / hv$, where *Bi* is the absorption constant for indirect transitions, *hv* is the photon energy, and E_g is the band gap energy [21]. Plots of $(ahv)^{1/2}$ versus *hv* from the spectral data are presented in Figure 3a, which shows the UV-visible spectra of the undoped (TiO_2) and $Fe/N-TiO_2$ particles from 250 to 700 nm. Samples A–E exhibited typical UV-visible spectra for semiconductor materials with a band gap absorption onset at 465, 388, 464, 452, and 485 nm, which corresponded to energy bandgaps at 2.67, 3.20, 2.67, 2.74, and 2.55 eV, respectively. These results demonstrate that the absorption of doped TiO_2 in the visible light region is significantly enhanced compared with that of undoped TiO_2, which in turn may considerably increase the photocatalytic activity of TiO_2 under visible light irradiation. Fourier-transform infrared (FT-IR) spectrum of the $Fe/N-TiO_2$ prepared using the sol-gel hydrothermal method at 150 °C and the undoped TiO_2 over the 400–4000 cm^{-1} range are shown in Figure 3b. The strong absorption at 3442 and 1640 cm^{-1} were assigned to the stretching vibration and the bending vibration of OH, respectively, originating from water adsorbed on the samples' surface [15]. The peaks around 1090 cm^{-1} were attributed to the N atoms embedded in the TiO_2 network. In addition, the small peak observed at 570 cm^{-1} indicates Fe–O–Ti vibrations [22]. No absorption peak for Fe–N stretching was observed, indicating that Fe did not substitute for Ti at sites where N atoms substituted for O atoms.

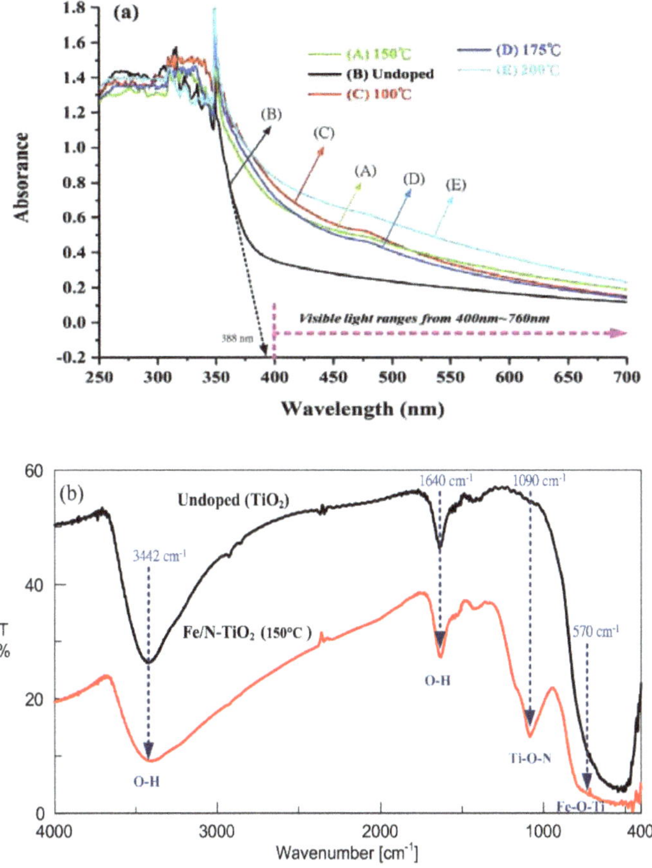

Figure 3. Fe/N-TiO$_2$ powders prepared using the sol-gel hydrothermal process: (**a**) UV-visible reflectance spectra of samples A–E, (**b**) FT-IR spectra of the samples.

The optimal synthesis temperature of Fe/N-TiO$_2$ was determined from batch experiments. Figure 4a shows the photocatalytic AO7 degradation curves for Fe/N-TiO$_2$ catalysts synthesized at different temperatures (see Table 1). The photocatalytic activity evolved as follows: Fe/N-TiO$_2$ (150 °C) > Fe/N-TiO$_2$ (175 °C) > Fe/N-TiO$_2$ (200 °C) > Fe/N-TiO$_2$ (100 °C) > undoped TiO$_2$. Fe/N TiO$_2$ (150 °C) exhibited the highest photocatalytic activity and led to 95.2% AO7 degradation in 5 h. In addition, Figure 4b plots ln(C/C_0) versus time obtained by assuming first-order kinetics for the degradation reaction. C and C_0 are the AO7 concentrations at time t and initial concentration, respectively. The plots were almost linear, indicating that the reactions followed pseudo first-order kinetics. The first-order degradation rate constants (k) for Fe/N-TiO$_2$ (150 °C), Fe/N-TiO$_2$ (175 °C), Fe/N-TiO$_2$ (200 °C), Fe/N-TiO$_2$ (100 °C), and undoped TiO$_2$ catalysts were 5.64 × 10^{-2}, 4.57 × 10^{-2}, 2.23 × 10^{-2}, 1.36 × 10^{-2}, and 8.53 × 10^{-1} min^{-1}, respectively. This suggests that codoping of Fe and N narrows the TiO$_2$ band gap. Cong et al. [23] reported that the overlap of the Ti-d orbital of TiO$_2$ and the doped metal d orbital leads to a narrowing of the TiO$_2$ band gap in TiO$_2$ implanted with metal ions, allowing the absorption of visible light. Therefore, N and Fe were incorporated into the TiO$_2$ framework, narrowing the band gap of TiO$_2$ to 2.67 eV (Table 1) and causing a large red shift, which in turn caused a much narrower band gap and greatly improved photocatalytic activity. By contrast, it inhibits the recombination of photogenerated electrons and holes. Fe ions with a suitable concentration can trap photogenerated electrons, which enhances the utilization efficiency of the photogenerated electron and hole [24]. Consequently, under these experimental conditions, Fe/N-TiO$_2$ (150 °C) was optimal for AO7 removal after 5 h of visible light irradiation time.

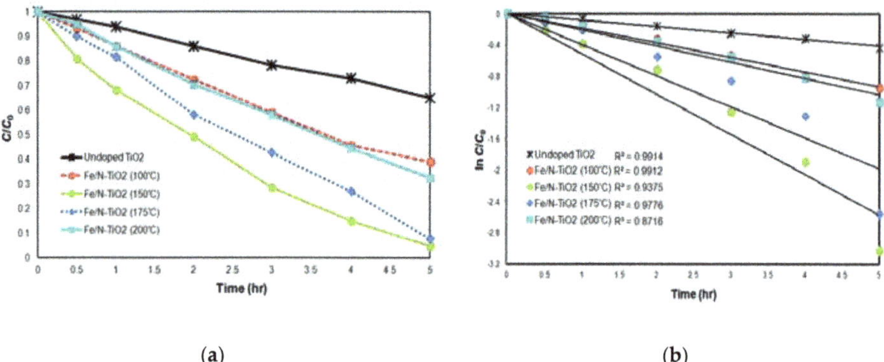

Figure 4. (a) AO7 degradation curves for Fe/N-TiO$_2$ catalysts synthesized at different temperatures; (b) logarithmic AO7 decay as a function of time. (Experimental condition: pH = 3, initial AO7 concentration = 10 mg/L, catalyst dosage = 0.1 g/L).

3.2. Degradation of AO7 in a Continuous-Flow Fluidized-Bed System

The optimal Fe/N-TiO$_2$ (150 °C) catalyst was selected for photocatalytic activity tests of the degradation of AO7. The effect of the initial AO7 concentration on the photocatalytic degradation efficiency was examined for concentrations ranging from 100 to 300 mg/L with an Fe/N-TiO$_2$ (150 °C) dosage of 20 g/L, a hydraulic retention time (HRT) of 10 h, a pH of 3, and a visible light intensity of 35 mW/cm^2. Figure 5a shows the AO7 removal efficiency and observed rate constant (K_{obs}) as a function of the initial AO7 concentration at a pH of 3 and with an HRT of 10 h. The degradation rate of AO7 decreased when the initial AO7 concentration increased. The number of photons decreased because of the decreasing intensity of the visible light, leading to a decrease in the formation of hydroxyl radicals, which ultimately reduced AO7 removal efficiency [25]. Moreover, the reaction rate also increased when the visible light intensity increased, and the maximum rate was reached for an irradiation of 35 mW/cm^2, as illustrated in Figure 5b. This indicates that the rate of photons per unit area of catalyst powder increased with the light intensity [26], and there was a corresponding increase in photocatalytic degradation rate of AO7.

To study the effect of pH on degradation efficiency, experiments were performed under visible light at pH values from 3 to 10 with constant concentrations of AO7 and Fe/N-TiO$_2$ (150 °C) catalyst. The results in Figure 6a indicate that the photodegradation efficiency for AO7 increased as the pH decreased, with maximum efficiency (88%) at pH 3. The degradation rates for the continuous-flow photoreactor evolved as follows: pH 3 > pH 7 > pH 10. In addition, increasing the HRT from 2.5 to 10 h increased the AO7 removal efficiency from 32% to 88% at pH 3. Explaining the effect of pH on the dye photodegradation efficiency is difficult because of the multiple roles of H$^+$ ions, but pH change is related to the charge in the functionalized surface of the solid catalyst according to the following reactions [27]:

$$\text{TiOH} + \text{H}^+ \longleftrightarrow \text{TiOH}_2{}^+, \qquad \text{pH} < \text{pH}_\zeta \tag{1}$$

$$\text{TiOH} + \text{OH}^- \longleftrightarrow \text{TiO}^- + \text{H}_2\text{O}, \qquad \text{pH} > \text{pH}_\zeta \tag{2}$$

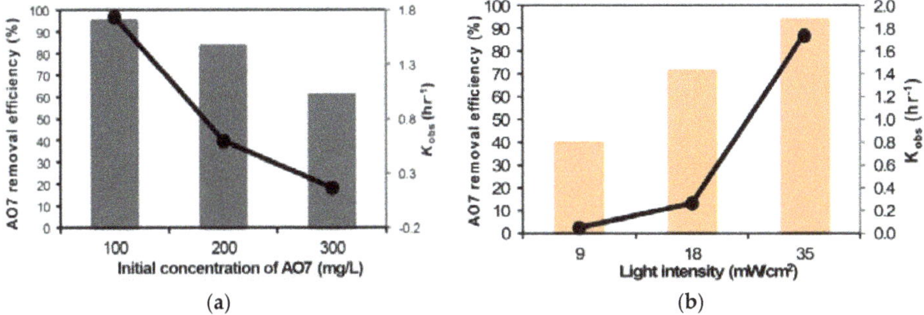

Figure 5. (**a**) AO7 removal efficiency and K_{obs} as a function of initial concentration, (**b**) effect of visible light intensity on AO7 removal efficiency and K_{obs} (Experimental condition: pH = 3, HRT = 10 h, catalyst dosage = 20 g/L).

According to Equation (1), when TiO_2 is suspended in an acidic solution (pH < point of zero charge, pH_ζ), the surface charge of TiO_2 becomes positive. Conversely, when TiO_2 is suspended in a basic solution (pH > pH_ζ), the surface charge becomes negative, as shown in Equation (2). Figure 6b shows that pH_ζ for the Fe/N-TiO$_2$ was 6. Therefore, the surface of the catalyst was positively charged at pH < 6 and negatively charged at pH > 6. AO7 is an anionic dye and was negatively charged under the experimental conditions used because of the SO_3^{2-} groups. Therefore, electrostatic interactions between the Fe/N-TiO$_2$ catalysts and the sulfonate groups resulted in adsorption at pH < 6 and enhanced degradation efficiency. Conversely, adsorption of AO7 onto Fe/N-TiO$_2$ surfaces was weak at pH > 6 because of Coulombic repulsion between the negatively charged Fe/N-TiO$_2$ and the AO7 molecules. Therefore, the degradation efficiency decreased.

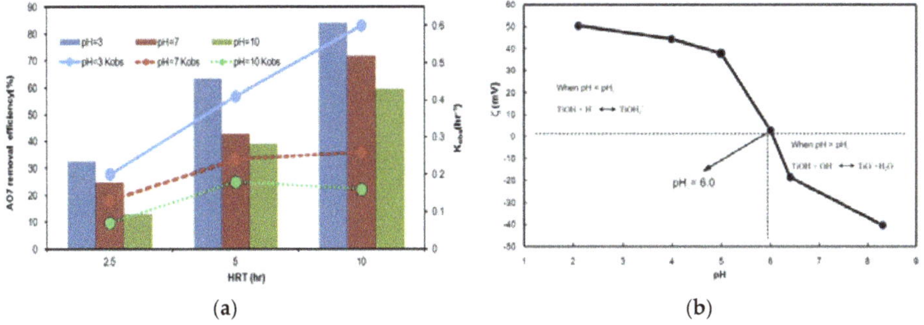

Figure 6. (**a**) Effect of pH on AO7 removal efficiency and K_{obs} as a function of HRT, (**b**) Zeta potential (ζ) of Fe/N-TiO$_2$ (initial AO7 concentration = 200 mg/L, catalyst dosage = 20 g/L, visible light intensity = 35mW/cm^2).

4. Conclusions

Fe/N-TiO$_2$ catalysts were synthesized using a combination of sol-gel and hydrothermal processes. The average size and distribution of the Fe/N-TiO$_2$ particles synthesized was approximately 10–15 nm. The average size of the particles synthesized increased with the reaction temperature, and the absorption edge of Fe/N-TiO$_2$ catalysts was red-shifted toward 480 nm. The Fe/N-TiO$_2$ photocatalyst exhibited favorable photocatalytic activity for the degradation of AO7 in a continuous-flow fluidized-bed system under visible light. The experimental results revealed that the optimal dosage of Fe/N-TiO$_2$ was 20 g/L, and that AO7 degradation efficiency was substantially improved by increasing

HRT from 2.5 to 10 h or by reducing initial AO7 concentration from 300 to 100 mg/L. Additionally, the degradation efficiency of AO7 increased as the pH decreased, with a maximum efficiency at pH 3.

Author Contributions: All the authors have contributed equally to the realization of work.

Funding: This study was supported by grant from Hsuteng Consulting International Co., Ltd.

Acknowledgments: The authors gratefully acknowledge experimental apparatus support from Institute of Environment Engineering and Management, National Taipei University of Technology.

Conflicts of Interest: The authors declare no conflicts of interest.

References

1. Easton, J.R.; Waters, B.D.; Churchley, J.H.; Harrison, J. Colour in dyehouse effluent. In *Society of Dyers and Colourists*; Cooper, P., Ed.; The Alden Press: Oxford, UK, 1995.

2. Liu, C.C.; Hsieh, Y.H.; Lai, P.F.; Li, C.H.; Kao, C.L. Photodegradation treatment of azo dye wastewater by UV/TiO$_2$ process. *Dyes Pigm.* **2006**, *68*, 191–195. [CrossRef]

3. Lu, X.; Yang, B.; Chen, J.; Sun, R. Treatment of wastewater containing azo dye reactive brilliant red X-3B using sequential ozonation and upflow biological aerated filter process. *J. Hazard. Mater.* **2009**, *161*, 241–245. [CrossRef] [PubMed]

4. Jothiramalingam, R.; Wang, M.K. Synthesis, characterization and photocatalytic activity of porous manganese oxide doped titania for toluene decomposition. *J. Hazard. Mater.* **2007**, *147*, 562–569. [CrossRef] [PubMed]

5. American Society for Testing and Material. *Powder Diffraction Files*; Joint Committee on Powder Diffraction Standards: Swarthmore, PA, USA, 1999; pp. 3–888.

6. McManamon, C.; Delaney, P.; Morris, M.A. Photocatalytic properties of metal and non-metal doped novel sub 10 nm titanium dioxide nanoparticles on methyl orange. *J. Colloid Interface Sci.* **2013**, *411*, 169–172. [CrossRef] [PubMed]

7. Irie, H.; Watanabe, Y.; Hashimoto, K. Nitrogen-concentration dependence on photocatalytic activity of TiO$_{2-x}$N$_x$ powders. *J. Phys. Chem. B* **2003**, *107*, 5483–5486. [CrossRef]

8. Lin, Z.; Orlov, A.; Lambert, R.M.; Payne, M.C. New Insights into the origin of visible light photocatalytic activity of nitrogen-foped and oxygen-feficient snatase TiO$_2$. *J. Phys. Chem. B* **2005**, *109*, 20948–20952. [CrossRef] [PubMed]

9. Zhou, M.; Yu, J.; Cheng, B. Effects of Fe-doping on the photocatalytic activity of mesoporous TiO$_2$ powders prepared by an ultrasonic method. *J. Hazard. Mater.* **2006**, *137*, 1838–1847. [CrossRef] [PubMed]

10. Yu, J.; Xiang, Q.; Zhou, M. Preparation, characterization and visible-light-driven photocatalytic activity of Fe-doped titania nanorods and first-principles study for electronic structures. *Appl. Catal. B* **2009**, *90*, 595–602. [CrossRef]

11. Wen, L.; Liu, B.; Zhao, X.; Kazuya, N.; Taketoshi, M.; Akira, F. Synthesis, characterization, and photocatalysis of Fe-doped TiO$_2$: A combined experimental and theoretical study. *Int. J. Photoenergy* **2012**, *2012*, 1–10.

12. Dang, T.M.H.; Trinh, V.D.; Bui, D.H.; Phan, M.H.; Huynh, D.C. Sol-gel hydrothermal synthesis of strontium hexaferrite nanoparticles and the relation between their crystal structure and high coercivity properties. *Adv. Nat. Sci. Nanosci. Nanotechnol.* **2012**, *2*, 1–7. [CrossRef]

13. Kim, H.J.; Bae, S.B.; Bae, D.S. Synthesis and characterization of Ru doped TiO$_2$ nanoparticles by a sol-gel and a hydrothermal process. *J Mater.* **2010**, *9*, 9–12.

14. Bickley, R.I.; Gonzalez-Carreno, T.; Lees, J.S.; Palmisano, L.; Tilley, R.J.D. A structural investigation of titanium dioxide photocatalysts. *J. Solid State Chem.* **1991**, *92*, 178–190. [CrossRef]

15. Kim, T.H.; Rodríguez-González, V.; Gyawali, G.; Cho, S.H.; Sekino, T.; Lee, S.W. Synthesis of solar light responsive Fe, N codoped TiO$_2$ photocatalyst by sonochemical method. *Catal. Today* **2013**, *212*, 75–80. [CrossRef]

16. Deng, L.; Wang, S.; Liu, D. Synthesis, characterization of Fe-doped TiO$_2$ nanotubes with high photocatalytic activity. *Catal. Lett.* **2009**, *129*, 513–518. [CrossRef]

17. Cheng, H.H.; Chen, S.S.; Cheng, Y.W.; Tseng, W.L.; Wang, Y.H. Liquid-phase non-thermal plasma-prepared N-doped TiO$_2$ for azo dye degradation with the catalyst separation system by ceramic membranes. *Water Sci. Technol.* **2010**, *62*, 1060–1066. [CrossRef] [PubMed]

18. Yang, G.; Jiang, Z.; Shi, H.; Xiao, T.; Yan, Z. Preparation of highly visible-light active N-doped TiO$_2$ photocatalyst. *J. Mater. Chem.* **2010**, *20*, 5301–5309. [CrossRef]

19. Hsu, B.C.; Chen, S.S.; Su, C.C.; Li, Y.C. Preparation and characterization of nanocrystalline Fe/N Codoped titania. *Ferroelectrics* **2009**, *381*, 51–58. [CrossRef]

20. Saha, N.C.; Saha, H.G. Titanium nitride oxidation chemistry: An X-ray photoelectron spectroscopy study. *J. Appl. Phys.* **1992**, *72*, 3072–3079. [CrossRef]

21. Tian, F.; Wu, Z.; Tong, Y.; Wu, Z.; Cravotto, G. Microwave-assisted synthesis of carbon-based (N, Fe)-codoped TiO$_2$ for the photocatalytic degradation of formaldehyde. *Nanoscale Res. Lett.* **2010**, *10*, 1–12. [CrossRef] [PubMed]

22. Ishii, M.; Nakahira, M.; Yamanaka, T. Infrared absorption spectra and cation distributions in (Mn, Fe)$_3$O$_4$. *Solid State Commun.* **1972**, *1*, 209–212. [CrossRef]

23. Cong, Y.; Zhang, J.; Chen, F.; Anpo, M.; He, D. Preparation, photocatalytic activity, and mechanism of nano-TiO$_2$ Codoped with nitrogen and iron (III). *J. Phys. Chem. C* **2007**, *111*, 10618–10623. [CrossRef]

24. Nguyen, V.N.; Nguyen, N.K.T.; Nguyen, P.H. Hydrothermal synthesis of Fe-doped TiO$_2$ nanostructure photocatalys. *Adv. Nat. Sci. Nanosci. Nanotechnol.* **2011**, *2*, 1–4.

25. Daneshvar, N.; Aber, S.; Hosseinzadeh, F. Study of C.I. acid orange 7 removal in contaminated water by photo oxidation processes. *Glob. Nest J.* **2008**, *10*, 16–23.

26. Sharma, R.; Heda, L.C.; Ameta, S.C.; Sharma, S. Use of zinc oxide as photocatalyst for photodegradation of copper soap derived from azadirecta indica (neem) oil. *Res. J. Pharm. Biol. Chem. Sci.* **2013**, *4*, 537–548.

27. Wang, W.Y.; Ku, Y. Effect of solution pH on the adsorption and photocatalytic reaction behaviors of dyes using TiO$_2$ and Nafion-coated TiO$_2$. *Colloid Surf. A Physicochem. Eng. Asp.* **2007**, *302*, 261–268. [CrossRef]

© 2018 by the authors. Licensee MDPI, Basel, Switzerland. This article is an open access article distributed under the terms and conditions of the Creative Commons Attribution (CC BY) license (http://creativecommons.org/licenses/by/4.0/).

 *materials*

Article

Tetranuclear Oxo-Titanium Clusters with Different Carboxylate Aromatic Ligands: Optical Properties, DFT Calculations, and Photoactivity

Maciej Janek [1], Aleksandra Radtke [1,2], Tadeusz M. Muzioł [1], Maria Jerzykiewicz [3] and Piotr Piszczek [1,2,*]

[1] Faculty of Chemistry, Nicolaus Copernicus University in Toruń, ul. Gagarina 7, 87-100 Toruń, Poland; maciejjanekk@gmail.com (M.J.); Aleksandra.Radtke@umk.pl (A.R.); tadeuszmuziol@wp.pl (T.M.M.)
[2] Nano-implant Ltd. Gagarina 5/102, 87-100 Toruń, Poland, NIP 9562314777
[3] Faculty of Chemistry, Wrocław University, ul. F. Joliot-Curie 14, 50-383 Wrocław, Poland; maria.jerzykiewicz@chem.uni.wroc.pl
*   Correspondence: piszczek@chem.umk.pl; Tel.: +48-56-611-45-92

Received: 26 July 2018; Accepted: 4 September 2018; Published: 8 September 2018

Abstract: Titanium(IV) oxo-clusters of the general formula $(Ti_4O_2(O^iBu)_{10}(O_2CR')_2)$ (R' = $C_{13}H_9$ (**1**), PhCl (**2**), PhNO$_2$ (**3**)) were studied in order to estimate their potential photoactivity. The structure of the resulting tetranuclear Ti(IV) oxo-complexes was then determined via single crystal X-ray diffraction, infrared and Raman spectroscopy, and electron spin resonance (ESR). An analysis of diffuse reflectance spectra (DRS) allowed for the assessment of band gap values of (**1**)–(**3**) microcrystalline samples complexes. The use of different carboxylate ligands allowed the band gap of tetranuclear Ti(IV) oxo-clusters to be modulated in the range of 3.6 eV–2.5 eV. Density functional theory (DFT) methods were used to explain the influence of substitutes on band gap and optical activity. Dispersion of (**1**)–(**3**) microcrystals in the poly(methyl methacrylate) (PMMA) matrixes enabled the formation of composite materials for which the potential photocatalytic activity was estimated through the study on methylene blue (MB) photodegradation processes in the presence of UV light. The results obtained revealed a significant influence of carboxylate ligands functionalization on the photoactivity of synthesized tetranuclear Ti(IV) oxo-complexes.

Keywords: titanium(IV) oxo-clusters; photoactivity; band gap modification; photoluminescence; DFT calculations; composite materials

---

1. Introduction

The unique physicochemical and biological properties of titanium dioxide favor its wide application in a variety of fields in our lives. The optical properties and photocatalytic activity of materials based of TiO$_2$ have especially been intensively studied in recent times [1–3]. Titania photoactivity is utilized in water splitting, purification of air and water, reduction of environmental pollutants, and in antimicrobial applications [4,5]. Recently, much attention has been devoted to the use of titanium oxo-clusters (TOCs) as compounds exhibiting similar properties to TiO$_2$ but characterized by a discrete molecular structure [6–10]. An analysis of the literature indicates the significance of TOCs in synthesis of inorganic–organic hybrid materials that are produced through introduction of metal oxo-clusters into the polymer matrix [11]. The possible interactions between inorganic and organic components may result in an improvement of structural properties of the polymer as well as its thermal, mechanical, and barrier properties due to cross-linking and filling. The unique properties of oxo-clusters, e.g., photochromicity, catalytic/biological or magnetic activity, can give entirely new properties to the composite material compared to the base polymer [11–22]. Therefore, studies on

TOC synthesis of the titanium oxide core with the desirable architecture, size, and physicochemical properties are important for the production of novel inorganic–organic composite materials [17].

A good example of the above is the studies on TOCs, which are stabilized by phosphonate and different carboxylic acids used as photocatalysts in the process of water splitting and photodegradation of organic dyes [22]. According to these investigations, the structural functionalization of carboxylate ligands associated with the size modification of the band gap had a direct impact on the photocatalytic activity of oxo-clusters. In addition, the use of $(Ti_6O_6(O^iPr)_6(O_2CR')_6)$ $(R' = C_6H_4NH_2, C_6H_4NHMe, C_6H_4NMe_2, C_6H_3(F)NH_2, C_6H_3(Cl)NH_2)$ complexes in the photocatalytical degradation of methylene blue (MB) and rhodamine B (RB) revealed a clear influence of the functionalization method of carboxyl groups on the energy band gap size and photocatalytic activity of these compounds [23]. Lin et al. described three TOCs with $\{Ti_6O_4\}$ core consisting of pivalic and benzoate ligands as photocatalysts bearing better water splitting properties than pure TiO_2 nanowires, nanotubes, and nanosheets phases [24]. Application of phthalic acid in synthesis of TOCs led to the formation of the oxo-complex, which contain $\{Ti_6O_3\}$ cores, which was used in the degradation of methyl orange in the presence of H_2O_2 [25]. In addition, $\{Ti_6O_3\}$ clusters with malonate and succinate ligands were studied for photodegradation of methyl orange [26].

In our previous works, we have focused on synthesis, structure determination, and photocatalytic activity studies of tetranuclear oxo-complexes [19,27]. The results of earlier studies revealed that three different types of $(Ti_4O_b(OR)_c(O_2CR')_{16-2b-c})$ complexes are usually formed in a direct reaction of $Ti(OR)_4$ and organic acid in the different alkoxide/acid molar ratio, i.e., (a) $\{Ti_4O_4\}$ of C.r. = 1 (C.r.—complexation ratio, e.g., $(Ti_4O_4(OR)_4(O_2CR')_4)$, R = ^iPr, ^tBu, R' = H, ^iPr, $C(Me)_2Et$, $Co_3C(CO)_9$ [19,21]); (b) $\{Ti_4O_2\}$ of C.r. = 1.5 (e.g., $(Ti_4O_2(OR)_6(O_2CR')_6)$, R = ^iPr, R' = C_2H_3, $C(Me)$ = CH_2, C_5H_4FeCp [21]); and (c) $\{Ti_4O_2\}$ of C.r. = 0.5 (e.g., $(Ti_4O_2(OR)_{10}(O_2CR')_2)$, R = ^iPr, ^iBu, R' = H, $C_6H_4NH_2$ [20,21,27]). Moreover, Liu et al. proved the formation of $(Ti_4O_3(O^iPr)_6(fdc)_2)$ complex, which consists of square-planar $\{Ti_4(\mu_4\text{-}O)(\mu\text{-}O)_2\}$ cores (C.r. = 0.75) in the hydrothermal conditions (the reaction of $Ti(O^iPr)_4$ with 1,1-ferrocenedicarboxylate acid ($fdcH_2$) in DMF) [28]. The structures of Ti(IV) oxo-complexes containing $\{Ti_4O\}$ cores (C.r. = 0.25) were also found, e.g., $(Ti_4O(OEt)_{12}(R'))$ (C.r. = 0.25, R' = O_3P-Phen, $^iBuPO_3$), $(Ti_4(\mu_3\text{-}O)(O^iPr)_8(O_3P\text{-}^tBu)_3(DMSO))$ [20].

From the abovementioned tetranuclear systems, the $(Ti_4O_2(O^iBu)_{10}(OOCC_6H_4NH_2)_2)$ was especially interesting for us. This complex was synthesized with a good yield in a direct reaction of $Ti(O^iBu)_4$ with 4-aminobenzoic acid in molar ratio 4:1 at room temperature [27]. The isolated crystals revealed hydrophobic properties and significantly lower sensitivity to moisture in comparison to compounds such as $(Ti_4O_4(OR)_4(O_2CC(Me)_2Et)_4)$ or $(Ti_6O_6(OR)_6(O_2CC(Me)_2Et)_6)$ (R = ^iBu, ^tBu) [19]. Moreover, the $(Ti_4O_2(O^iBu)_{10}(O_2CC_6H_4NH_2)_2)$ compound revealed a promising activity in photocatalytic MB degradation under UV light irradiation. Having considered previous reports on the influence of the type of carboxylate ligand on the optical properties and photocatalytic activity of oxo-clusters, we decided on the synthesis of new $(Ti_4O_2(O^iBu)_{10}(O_2CR')_2)$ complexes with different carboxylate ligands, i.e., fluorene-9-carboxylate ($O_2CC_{13}H_9$), 3-chlorobenzoic carboxylate ($O_2CC_6H_4Cl$), and 3-nitrobenzoic carboxylate ($O_2CC_6H_4NO_2$). Besides synthesis of new tetranuclear Ti(IV) oxo-complexes and their structural and spectral characteristic, it was interesting to determine the impact of carboxylate ligands on the $\{Ti_4O_2\}$ core structure, band gap, optic properties, and photocatalytic activity.

2. Materials and Methods

2.1. General Information

Titanium(IV) isobutoxide (Aldrich, St. Louis, MO, USA), 3-chlorobenzoic acid (Aldrich), 3-nitrobenzoic acid (Aldrich), 4-aminobenzoic acid (Aldrich), and fluorine-9-coarboxylic acid (Organic Acros, Geel, Belgium) were purchased commercially and were used without purification. The solvents applied—toluene, tetrahydrofuran, and acetone—were distilled before their use. All operations were performed in an argon atmosphere using standard Schlenk techniques. The vibrational spectra of synthesized compounds were registered using the Perkin Elmer Spectrum 2000 FT-IR spectrometer (400–4000 cm^{-1} range, KBr pellets) and the RamanMicro 200 spectrometer (PerkinElmer, Waltham, MA, USA). ^{13}C NMR spectra were carried out on Bruker Advance 700 (Madison, WI, USA). Elemental analyses were performed on Elemental Analyzer vario Macro CHN Elementar Analysensysteme GmbH (Langenselbold, Germany). The solid state optical diffuse reflection experiment was carried out on the Jasco V-750 Spectrophotometer (JASCO Deutschland GmbH, Pfungstadt, Germany) equipped with an integrating sphere for diffuse reflectance spectroscopy. The recorded spectra were analyzed using Spectra Manager^{TM} CFR software (JASCO Deutschland GmbH, Pfungstadt, Germany). The solid-state luminescence spectra were recorded on Gilden λ Photonics (Glasgow, UK). All ESR spectra were recorded at room temperature using a Bruker Elexsys E500 spectrometer (Rheinstetten, Germany) at a microwave power of 20 mW and modulation amplitude of 0.5 G. In the case of low intensity signals, five scans were accumulated. The measurements were performed for powdered samples immobilized on poly(methyl methacrylate) (PMMA) foils. Samples on foils were studied using special EPR Tissue Sample Cells (Wilmad-Labglass, WG-807 type, Vineland, NJ, USA). All samples were irradiated in situ using the 100 W mercury lamp (Bruker ER 203UV Irradiation System, Rheinstetten, Germany). To determine the diagonal components of g-tensor programs, SimFonia 1.26 developed by Bruker was employed.

2.2. Syntheses

The synthesis of (Ti$_4$O$_2$(O^{i}Bu)$_{10}$(O$_2$CC$_{13}$H$_9$)$_2$) (**1**): 0.184 g of 9-fluorenecarboxylic acid (0.875 mmol) was added to the solution of 1.19 g titanium(IV) isobutoxide (3.5 mmol) in 2 mL of acetone. Reactants underwent rapid reaction, leading to clear brownish solution. The solution was left for crystallization. Slow evaporation under an inert gas atmosphere led to crystals suitable for X-ray diffraction experiment. The yield basing on acid: 77% (0.46 g). Anal. Calcd. for C$_{68}$H$_{108}$O$_{16}$Ti$_4$: C, 59.28; H, 7.93; Ti, 13.94. Found: C, 58.92; H, 7.81; Ti, 13.79.

The synthesis of (Ti$_4$O$_2$(O^{i}Bu)$_{10}$(O$_2$CC$_6$H$_4$Cl)$_2$) (**2**): 0.137 g of 3-chlorobenzoic acid (0.875 mmol) was added to the solution of 1.19 g titanium(IV) isobutoxide (3.5 mmol) in 2 mL of toluene, leading to a colorless solution. The solution was left for crystallization. Slow evaporation under an inert gas atmosphere led to crystals suitable for X-ray diffraction experiment. The yield basing on acid: 88% (0.49 g). Anal. Calcd. for C$_{54}$H$_{98}$O$_{16}$Cl$_2$Ti$_4$: C, 51.24; H, 7.80; Ti, 15.13. Found: C, 50.89; H, 7.98; Ti, 14.95.

The synthesis of (Ti$_4$O$_2$(O^{i}Bu)$_{10}$(O$_2$CC$_6$H$_4$NO$_2$)$_2$) (**3**): 0.146 g of 3-nitrobenzoic acid (0.875 mmol) was added to the solution of 1.19 g titanium(IV) isobutoxide (3.5 mmol) in 2 mL of THF, leading to a weak yellow solution. The solution was left for crystallization. Slow evaporation under an inert gas atmosphere led to crystals suitable for X-ray diffraction experiment. The yield basing on acid: 94% (0.53 g). Anal. Calcd. for C$_{54}$H$_{98}$O$_{20}$N$_2$Ti$_4$: C, 50.40; H, 7.68; N, 2.18; Ti, 14.88. Found: 50.94; H, 7.74; N, 2.22; Ti, 14.71.

The synthesis of (Ti$_4$O$_2$(O^{i}Bu)$_{10}$(O$_2$CC$_6$H$_4$NH$_2$)$_2$) (**4**): Complex of a given formula was synthesized, as reported [27].

2.3. X-ray Crystalography Study

For single crystals, the diffraction data of (**1**) and (**2**) were collected using BL14.3 beamline (Helmholtz Zentrum Berlin, Germany, Bessy II), radiation λ = 0.89429 Å, at liquid nitrogen temperature, whereas for (**3**), the diffraction experiment was performed at room temperature using Oxford Sapphire CCD diffractometer, MoKα radiation λ = 0.71073 Å. The data were processed using CrysAlis [29], *xdsapp* [30], and XDS [31], and the numerical absorption correction was applied for all crystals. The structures of all complexes were solved by direct methods and refined with full-matrix least-squares procedure on F^2 (SHELX-97 [32]). All heavy atoms were refined with anisotropic displacement parameters. The positions of hydrogen atoms were assigned at calculated positions with thermal displacement parameters fixed to a value of 20% or 50% higher than those of the corresponding carbon atoms. For (**2**) and (**3**), some constraints (DFIX, DANG and ISOR) were applied for C–Cl, C–O, and C–C bonds. In (**1**) the alternate positions were found only for one $–O^iBu$ (O61), whereas for (**2**) a positional disorder was detected for chlorine atoms in O201, O211, and O221 carboxylate anions and in O81, O131 and O171 $–O^iBu$ ligands. In (**3**), we observed a significant disorder of O71, O91, O101, O121, O131, O141 and O191 $–O^iBu$. In the final model of (**3**), O141 isobutanolate was incomplete—there was lack of carbon and hydrogen atoms from minor orientation. All figures were prepared in DIAMOND [33] and ORTEP-3 [34]. The results of the data collections and refinement are summarized in Table 1. The crystallographic data of the complex (**4**) were presented in our earlier paper [27].

Table 1. The selected crystal data and structure refinements for $(Ti_4O_2(O^iBu)_{10}(O_2CR')_2)$ (R′ = $C_{13}H_9$ (**1**), PhCl (**2**), PhNO$_2$ (**3**); the complete crystallographic data are given in Table S1).

Parameters	(1)	(2)	(3)
Empirical formula	$C_{68}H_{108}O_{16}Ti_4$	$C_{54}H_{98}Cl_2O_{16}Ti_4$	$C_{54}H_{98}N_2O_{20}Ti_4$
Formula weight (g/mol)	1373.14	1265.82	1286.94
Temperature (K)	100(2)	100(2)	293(2)
Wavelength (Å)	0.89429	0.89429	0.71073
Space group	Monoclinic, P 1 21 1	Tetragonal, P 4 1	Triclinic, P 1
Unit cell dimensions (Å) and angles (°)	$a = 12.447(3)$ $b = 21.693(4)$ $c = 14.071(3)$ $\alpha = \gamma = 90$ $\beta = 105.19(3)$	$a = 17.827(3)$ $c = 41.475(8)$ $\alpha = \beta = \gamma = 90$	$a = 17.9713(9)$ $b = 18.0686(7)$ $c = 22.7006(9)$ $\alpha = 105.742(4)$ $\beta = 98.362(4)$ $\gamma = 92.631(4)$
Volume (Å^3)	3666.61(140)	13180.83(500)	6991.1(5)
Z, Calculated density (Mg/m^3)	2, 1.244	8, 1.276	4, 1.223
Final R indices (I > 2sigma(I))	R_1 [a] = 0.0444, wR_2 [b] = 0.1182	R_1 [a] = 0.0680, wR_2 [b] = 0.1851	R_1 [a] = 0.0860, wR_2 [b] = 0.2304
Absolute structure parameter	0.041(3)	0.387(19)	N/A

[a] $R_1 = \Sigma||F_0| - |F_c||/\Sigma|F_0|$; [b] $wR_2 = (\Sigma w(F_0^2 - F_c^2)^2/\Sigma(w(F_0^2)^2))^{1/2}$. CCDC 1826273, 1826276, and 1826277 contain the supplementary crystallographic data for (**1**), (**2**), and (**3**), respectively. These data can be obtained free of charge via http://www.ccdc.cam.ac.uk/conts/retrieving.html or from the Cambridge Crystallographic Data Centre, 12 Union Road, Cambridge CB2 1EZ, UK; fax: (+44) 1223-336-033; or e-mail: deposit@ccdc.cam.ac.uk.

2.4. Preparation and Photoactivity Study of Composites

Hybrid PMMA + {TOCs} composite foils containing 20 wt % of synthesized oxo-clusters were prepared by dissolving 1.0 g of polymer in 5 mL of THF and adding a solution of 0.25 g of TOCs—(**1**), (**2**), (**3**), and (**4**)—in 1 cm^3 of THF to this solution. The resulting mixtures were stirred, poured into a glass Petri dish and left for evaporation of the solvent. Then, the prepared polymer foils were characterized by Raman and NMR spectroscopy and scanning electron microscopy.

The photoactivity properties of polymer composite foils were studied through MB degradation. Foil samples of size 8 mm × 8 mm were placed in quartz cuvettes, and MB solution (V = 3 cm^3, initial concentration $C_0 = 1.0 \times 10^{-5}$ M) was poured on films. The prepared samples were exposed to UV irradiation (18 W, range of 340 nm–410 nm with maximum at 365 nm). MB absorbance at 660 nm was registered after 0, 24, 48, 72, 94, 168, and 192 h of irradiation. For the purpose of determination of photoactivity, the methodology of chemical kinetics assuming a Langmuir–Hinshelwood (L–H) reaction mechanism was used. However, the linear dependencies of $A = f(t)$ suggested that $Kc << 1$ in the L–H equation and it could be simplified. Thus, the rate of reaction can be expressed as:

$$r = -dc/dt = k_{deg}Kc/(1 + Kc) \approx k_{deg}Kc = k_{obs}c \tag{1}$$

where c is a reactant (MB) concentration, t is time of the concentration measurement, k_{deg} is the rate constant of MB degradation on the surface, k_{obs} is a first order observed rate constant, and K describes the reactant adsorption process.

2.5. The Computational Details

The density functional theory (DFT) calculations were performed with Gaussian09 packages [35]. The HSE06 functional was used with 6-31G(d,p) basis set. The crystal structures were used as an optimization starting geometry. In order to reduce the complexity of calculations, isobutyl moieties were substituted with methyl groups. The criterion of no imaginary frequencies was used to confirm the result as an actual minimum. Density-of states (DOS) plots were visualized with GaussSum 3.0 [36].

3. Results

3.1. Structures of $(Ti_4O_2(O^iBu)_{10}(O_2CR')_2)$ Clusters

Structural studies of (**1**)–(**3**) complexes proved that their central part was formed by $\{Ti_4O_2\}$ core that consisted of two μ-oxo bridges, i.e., μ_4–O connecting all four Ti(IV) atoms in distorted tetrahedral environment and μ–O bridging two of the closest titanium atoms (Figure 1) similar to the structure of (**4**) [27].

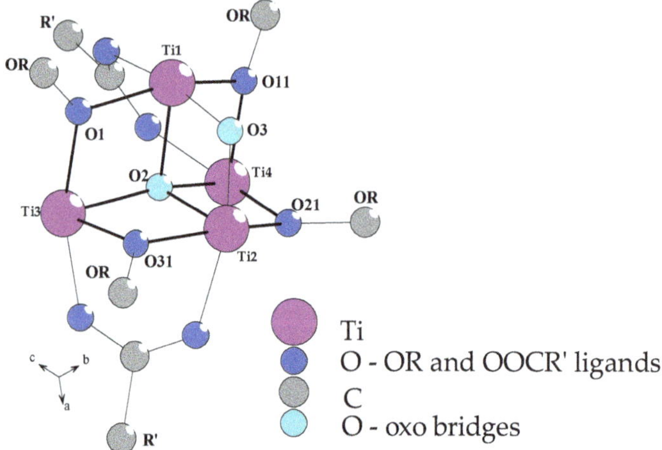

Figure 1. The structure of $\{Ti_4O_2\}$ core, which was found in $(Ti_4O_2(OR)_{10}(O_2CR')_2)$ (R = ^iBu, R' = $C_{13}H_9$ (**1**), PhCl (**2**), PhNO$_2$ (**3**)) complexes (crystallographic ball-stick scheme). For clarity, the terminal alkoxide groups are omitted.

Simultaneously, the $\{Ti_4O_2\}$ skeleton was stabilized by four μ_2-O^iBu bridges and two *syn–syn* carboxylate ligands (–O$_2$CR', R' = $C_{13}H_9$ (**1**), PhCl (**2**), PhNO$_2$ (**3**)). The selected bond lengths and angles, which were found in structures of (**1**)–(**3**) compounds, are listed in Table 2. Comparison of the structural data of these complexes and (**4**) [27] made it possible to trace the influence of carboxylate groups on the geometry of the $\{Ti_4O_2\}$ core, which may have a direct impact on their photocatalytic activity. According to earlier reports concerning photoactivity of Ti(IV) oxo-clusters, the possible changes of $\{Ti_2(\mu_2$–O)$\}$ bridges angle, which can be responsible for the facilitation of the photocatalytic process, should be noted [26]. In the group of synthesized compounds, the values of these angles changed in the row: 105.7(2) (**4**) [27] < 106.09(15) (**1**) < 106.5(2) (**2**) < 107.36(15) (**3**) (Table 2). It suggests that photoactivity of oxo-clusters should also change according to the above dependence. Changes of $\{Ti_2(\mu_2$–O)$\}$ bridges angle were associated with the differences in Ti1–Ti2 distances ((**1**) < (**2**) < (**3**) <

(**4**) (2.9521(13) [27]). At the same time, the geometry of {Ti$_2$(μ_4–O)} bridges also changed, which was reflected by differences in Ti3–Ti4 distances and Ti1–O2–Ti2 and Ti3–O2–Ti4 angles (Table 2).

Table 2. Selected bond lengths (Å) and angles (°) of (Ti$_4$O$_2$(O^{i}Bu)$_{10}$(O$_2$CR')$_2$) (R' = C$_{13}$H$_9$ (**1**), PhCl (**2**), PhNO$_2$ (**3**)).

Parameter		(**1**)	(**2**)	(**3**)
		Distances (Å)		
Ti–Ti	Ti1–Ti3	3.1690(11)	3.2016(16)	3.2060(12)
	Ti1–Ti2	2.9427(12)	2.9488(16)	2.9521(13)
	Ti1–Ti4	3.1446(12)	3.1456(15)	3.1508(12)
	Ti2–Ti3	3.1566(11)	3.1532(17)	3.1617(13)
	Ti2–Ti4	3.1738(13)	2.1859(17)	3.2009(12)
	Ti3–Ti4	4.0386(16)	3.9255(16)	3.9510(13)
Ti–(μ_4–O)	Ti1–O2	2.0457(30)	2.0599(39)	2.059(3)
	Ti2–O2	2.0357(28)	2.0622(41)	2.052(3)
	Ti3–O2	2.0820(27)	2.0603(38)	2.074(3)
	Ti4–O2	2.0896(27)	2.0264(38)	2.042(3)
Ti–(μ_2–O)	Ti1–O3	1.8326(37)	1.8643(40)	1.839(3)
	Ti3–O3	1.8499(41)	1.8182(39)	1.826(3)
Ti–(μ_2–OR)	Ti1–O11	1.9609(27)	1.9515(37)	1.944(3)
	Ti1–O1	2.0042(28)	2.0133(38)	2.017(3)
	Ti2–O21	1.9979(27)	2.0095(38)	2.020(3)
	Ti2–O31	1.9633(28)	1.9858(38)	1.969(3)
	Ti3–O1	1.9896(36)	2.0041(39)	1.993(3)
	Ti3–O31	2.0981(29)	2.0698(45)	2.089(3)
	Ti4–O11	2.0944(32)	2.1355(43)	2.114(3)
	Ti4–O21	1.9985(30)	2.0182(44)	2.004(3)
Ti–O (carb)	Ti4–O111	2.027(3)	2.051(4)	2.068(4)
	Ti1–O112	2.195(3)	2.135(4)	2.145(3)
	Ti2–O131	2.190(3)	2.127(5)	2.142(4)
	Ti3–O132	2.028(3)	2.073(5)	2.056(4)
O–C (carb)	O111–C112	1.270(5)	1.272(8)	1.267(6)
	O112–C112	1.245(5)	1.244(8)	1.240(6)
	O131–C132	1.245(6)	1.259(9)	1.260(7)
	O132–C132	1.257(6)	1.255(9)	1.245(6)
		Angles (°)		
Ti–(μ_4–O)–Ti	Ti3–O2–Ti2	100.10(12)	99.80(17)	100.05(12)
	Ti3–O2–Ti1	100.31(13)	102.02(17)	101.76(13)
	Ti2–O2–Ti1	92.28(12)	91.36(17)	91.79(12)
	Ti3–O2–Ti4	150.98(16)	147.7(2)	147.43(15)
	Ti2–O2–Ti4	100.58(13)	102.34(18)	102.83(14)
	Ti1–O2–Ti4	98.99(12)	100.66(18)	100.39(12)
Ti–(μ_2–O)–Ti	Ti1–O3–Ti2	106.09(15)	106.5(2)	107.36(16)
Ti–(μ_2–OR)–Ti	Ti1–O1–Ti3	105.04(15)	105.72(18)	106.16(14)
	Ti1–O11–Ti4	101.65(13)	100.52(19)	101.79(13)
	Ti2–O31–Ti3	101.97(14)	102.04(18)	102.35(13)
	Ti2–O21–Ti4	105.16(14)	104.55(19)	105.38(15)
O–C–O (carb)	O111–C112–O112	126.0(4)	125.5(6)	125.5(5)
	O131–C132–O132	126.6(4)	126.1(6)	125.8(5)

3.2. Analysis of Vibrational Spectra

The possible influence of the carboxylic groups on the structure of (Ti$_4$O$_2$(O^{i}Bu)$_{10}$(O$_2$CR')$_2$) clusters were studied using IR and Raman spectroscopy. The position of the bands derived from the vibrations of carboxylate ligands (ν_{as}(COO) and ν_s(COO) at ~1600 and ~1400 cm^{-1}) and {Ti$_4$O$_2$} cores (specific bridges: Ti(μ–O), Ti(μ_4–O) at 400–700 cm^{-1}) were especially important for us (Table 3).

Table 3. Results of vibrational spectra studies of **(1)**–**(3)** complexes.

Modes	(1)		(2)		(3)	
	IR	**R**	**IR**	**R**	**IR**	**R**
ν_{as}(COO)	1566 (s)	1611 (s) 1581 (w)	1590 (s) 1551 (m)	1595 (s) 1552 (m)	1595 (m) 1559 (s)	1614 (w) 1599 (vw) 1561 (m)
ν_s(COO)	1448 (m)	1450 (m)	1442 (m)	1459 (m)	1434 (m)	1456 (s)
ν_{as}(NO$_2$)	-	-	-	-	1530 (m)	1534 (m)
ν_s(NO$_2$)	-	-	-	-	1347 (s)	1349 (m)
ν_a(Ti–μ–O–Ti)	712 (w)	700 (m)	-	699 (m)	-	697 (m)
ν_s(Ti–μ–O–Ti)		666 (m)		661 (m)		653 (m)
ν_a(Ti–μ_4–O–Ti)	636 (m)	543 (vw)	635 (m)	548 (vw)	643 (m)	536 (vw)
ν_s(Ti–μ_4–O–Ti)	546 (s)	414 (m)	547 (m)	415 (m)	546 (s)	412 (m)

Earlier studies of coordination compounds with {Ti$_4$O$_2$} moiety revealed that the M–O stretches region of this type of compound is represented by many broad bands derived of different M–O modes between 400 cm^{-1} and 800 cm^{-1} [27,37]. The same pattern is shown for compounds bearing {Ti$_4$(μ_4–O)} core [38,39]. Our earlier investigations involving comparison of experimental spectra and DFT calculations of (Ti$_4$O$_2$(O^{i}Bu)$_{10}$(ABZ)$_2$) cluster (**4**) have revealed that asymmetric and symmetric Ti(μ–O) bridge bands are located between 700 cm^{-1} and 680 cm^{-1} and Ti(μ_4–O) in ~630 cm^{-1} and ~535 cm^{-1} [27]. Bands in 700–690 cm^{-1} come from symmetric ν_s(Ti–O–Ti) modes; ones at ~630 cm^{-1}, ~535 cm^{-1} and 430–420 cm^{-1}, which were found in IR and Raman spectra of this compound, can be assigned as well as ν_s(Ti–O–Ti) modes. DFT calculations (HSE06/6-31G(d,p) level of theory) were also carried out for the studied systems consisting of {Ti$_4$O$_2$} core linked with two carboxylate ligands—**(1)**–**(3)**—and stabilized by ten alkoxide groups (experimental and calculated structural data of {Ti$_4$O$_2$} cores are compared in Table S1). Results of these studies are summarized in Table 4. According to these data, the frequency of symmetric (s) and asymmetric (a) stretching modes of Ti–O–Ti bridges should be 699–706 cm^{-1} for μ–O bridges and 439–622 cm^{-1} for μ_4–O ones. The bands, which can be attributed to abovementioned modes, were also found in IR and Raman spectra of **(1)**–**(3)** complexes (Table 3, Figure S1). Analysis of spectral data exhibited that a significant effect of carboxylic groups on the vibration frequency of Ti–O–Ti bridges was not observed.

Table 4. The calculated frequencies of the (Ti–O–Ti) modes of {Ti$_2$-(μ_2-O)} and {Ti$_4$-(μ_4-O)} bridges. In DFT calculations of **(1)**–**(3)** complexes the O^{i}Bu ligands were exchanged on the OMe one.

Complex	Mode	Frequency (cm^{-1})
(Ti$_4$O$_2$(OMe)$_{10}$(O$_2$CC$_{13}$H$_9$)$_2$) **(1)**	ν_a(Ti–μ–O–Ti)	703
	ν_s(Ti–μ–O–Ti)	700
	ν_a(Ti–μ_4–O–Ti)	623
	ν_s(Ti–μ_4–O–Ti)	570,439
(Ti$_4$O$_2$(OMe)$_{10}$(O$_2$CC$_6$H$_4$Cl)$_2$) **(2)**	ν_a(Ti–μ–O–Ti)	705
	ν_s(Ti–μ–O–Ti)	700
	ν_a(Ti–μ_4–O–Ti)	622
	ν_s(Ti–μ_4–O–Ti)	560,448
(Ti$_4$O$_2$(OMe)$_{10}$(O$_2$CC$_6$H$_4$NO$_2$)$_2$) **(3)**	ν_a(Ti–μ–O–Ti)	706
	ν_s(Ti–μ–O–Ti)	699
	ν_a(Ti–μ_4–O–Ti)	622
	ν_s(Ti–μ_4–O–Ti)	567,444

3.3. UV-Vis Absorption Spectra, Band Gap Determination and DOS

The solid state UV-Vis absorption spectra of studied compounds were recorded with an integrating sphere fitted spectrophotometer at room temperature (Figure 2a). MgO was used as a standard reference. Figure 2b shows the plot of Kubelka–Munk (K–M) function versus light energy, i.e., K = f(hν) where K = $(1 - R)^2/2R$ and R is reflectance that was used for the optical band gap determination [40,41].

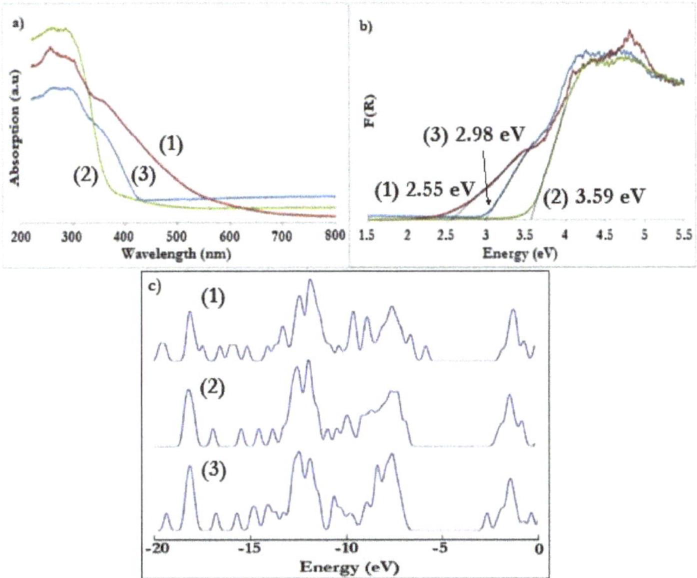

Figure 2. (a) Solid-state UV-Vis absorption spectra of (1)–(3) micrograins; (b) Kubelka–Munk function versus light energy plot used for band gap determination; (c) density-of states (DOS) plots calculated with HSE06/6-31G(d,p) level of theory for $(Ti_4O_2(OMe)_{10}(OOC-C_{13}H_9)_2)$ (1), $(Ti_4O_2(OMe)_{10}(OOC-C_6H_4-Cl)_2)$ (2), and $(Ti_4O_2(OMe)_{10}(OOC-C_6H_4-NO_2)_2)$ (3).

Compound (3) absorbed exclusively in UV region of the spectrum, which corresponded to band gap of 3.59 eV. According to theoretical calculations, the chloro-functionalized derivative (2) introduced additional band as HOMO, but it didn't change the value of resulting band gap by a big margin (Figure 2c, Figure S2). By contrast, DOS plots of nitro- and fluorene-derivative showed deep penetration of the gap by functionalities, and reduction of the band gap was therefore evidenced. It is noteworthy that 3-nitrobenzoic derivative (3) introduced new states as LUMO, while compound with fluorene (1) changed HOMO of the corresponding compound (Figure S3).

Theoretically calculated band gaps were qualitatively reflected in absorption spectra of (1) and (3), where the bands penetrated into visible region of spectrum (Table 5). The band gaps of (1) and (3) were estimated by K–M function onset to be 2.98 eV and 2.55 eV, respectively. Calculated and experimentally determined values of band gaps are presented in Table 5. There was a clear overestimation of the energy gap between the values determined experimentally and theoretically; however, this agrees with the tendency of hybrid density functionals to overestimate predicted band gaps [42]. Additionally, the comparison of experimental vs. theoretical values showed that they follow a linear trend (R^2 = 0.9968), indicating correlation between them. The band gap of (4) was determined in previous research to be 2.57 eV [27].

Table 5. Comparison of experimentally determined band gap values (using diffuse reflectance spectra), and theoretically calculated (HSE06/6-31(d,p) level of theory in calculations of (**1**)–(**3**) complexes the O^{i}Bu ligands were exchanged on the OMe one).

Complex	Calculated Band Gap (eV)	Experimental Band Gap (eV)
(Ti$_4$O$_2$(OMe)$_{10}$(O$_2$CC$_{13}$H$_9$)$_2$) (**1**)	3.93	2.55
(Ti$_4$O$_2$(OMe)$_{10}$(O$_2$CC$_6$H$_4$Cl)$_2$) (**2**)	4.73	3.59
(Ti$_4$O$_2$(OMe)$_{10}$(O$_2$CC$_6$H$_4$NO$_2$)$_2$) (**3**)	4.30	2.98

3.4. Photoluminescent Properties

Solid-state photoluminescence spectra of (**1**)–(**4**) were recorded with samples excited by wavelength of 330 nm (3.76 eV), i.e., energy higher than determined for band gaps (Figure 3). Resulting spectra were compared with the data of corresponding acids. Results showed how coordination to oxo-titanium cluster influenced ligands PL spectra. In the case of (**1**) and (**2**), the coordination to oxo-titanium core causes the decay of higher energy part of the corresponding acid spectrum, indicating interactions of the frontier orbitals of carboxylate with oxo-titanium core. The change in band location was seen in case of fluorene derivative (**2**) where the emission of fluorine-9-carboxylic acid at 540 nm shifted toward 560 nm when coordinated. This type of conversion was not witnessed for other ligands. Interestingly, photoluminescent spectra of (**3**) and (**4**) didn't show any distinct changes when compared to spectra of acids.

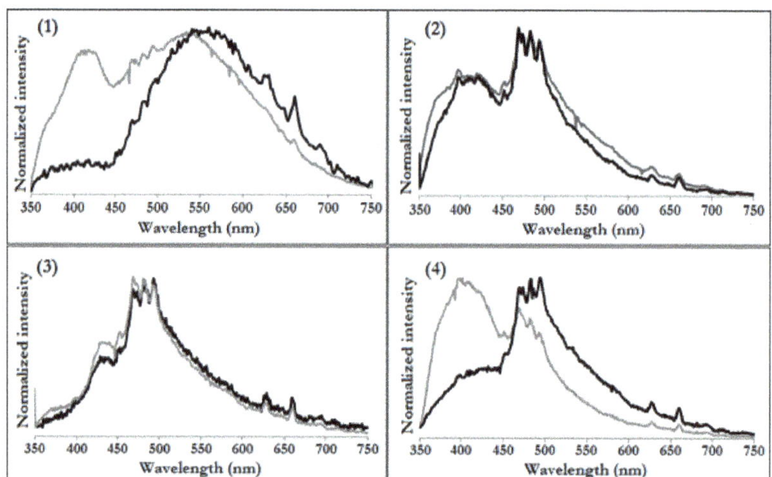

Figure 3. Solid state photoluminescence spectra (1–4) of (Ti$_4$O$_2$(O^{i}Bu)$_{10}$(O$_2$CR′)$_2$) (R′ = C$_{13}$H$_9$ (**1**), PhCl (**2**), PhNO$_2$ (**3**), PhNH$_2$ (**4**)) respectively (black plots) collected by 330 nm excitation at room temperature compared to spectra of corresponding acids (grey plots).

3.5. Preparation and Characteristic of Polymer Composites (PMMA + TOCs)

I In order to confirm the presence of the structurally stable Ti(IV) oxo-complexes in polymer, Raman spectra of composites were registered (Figure 4). Analysis of these data clearly indicates the presence of (**1**)–(**4**) complexes in the produced composite (PMMA + TOCs), which is evidenced by the appearance of bands attributed to the carboxylate ligands. Analysis of the region between 600 cm^{-1} and 750 cm^{-1}, in which appears the bands coming from the modes of oxo-titanium bridges, is difficult due to overlapping of ν(Ti–O–Ti) bands and poly(methyl methacrylate) ones. The presence of micrograins and nanograins of complexes in polymer matrices were also seen in SEM images of PMMA + TOCs foils (Figure 5).

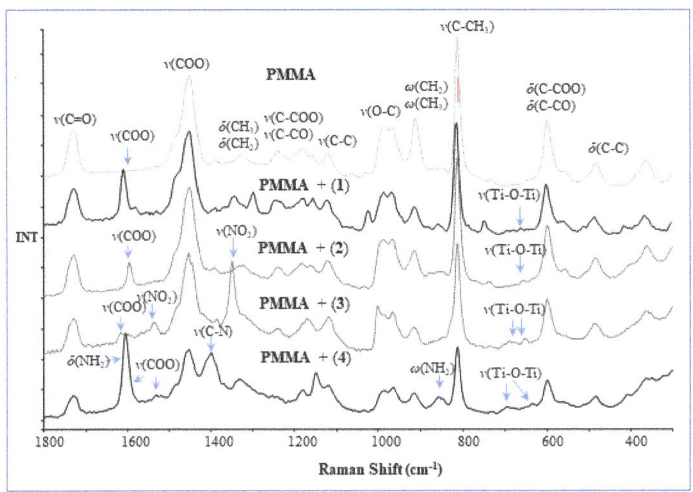

Figure 4. Raman spectra of poly(methyl methacrylate) (PMMA) and PMMA–titanium oxo-clusters (TOCs) composites. Arrows indicate the bands, derived from (1)–(4) complexes.

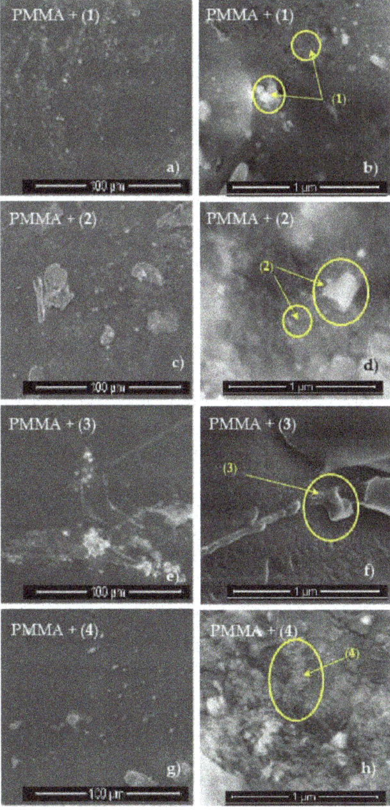

Figure 5. SEM images of PMMA + TOCs composites in different magnifications (1) (**a,b**), (2) (**c,d**), (3) (**e,f**), end (4) (**g,h**).

3.6. Photocatalytic Degradation of Methylene Blue (MB)

To evaluate the influence of the different functionalities on photoinduced behavior, the photocalatytic degradation of MB were executed. The degradation of MB on polymer composites was followed by UV-Vis spectrophotometry by monitoring changes in the absorbance peak at 660 nm. Kinetic tests involved behavior of the studied complexes as well as blind tests (unmodified PMMA foil and sole MB solution). The changes in absorbance were roughly linear (Figure 6) which can be resulted the relatively small PMMA foil surfaces.

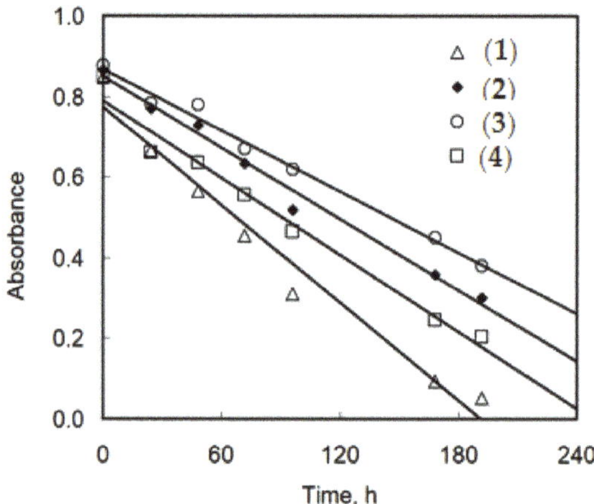

Figure 6. Changes in absorbance of methylene blue (MB) solution with time after addition of PMMA foil with a selected complex. MB = 1×10^{-5} M, PMMA foil surface = 0.64 cm^2, T = r.t., l = 1 cm. Presented values are after blind test correction (R^2 = 0.9701 (**1**), 0.9876 (**2**), 0.9896 (**3**), 0.9769 (**4**)).

Thus, the rate of MB degradation was constant as long as the MB concentration was relatively high. Such behavior meant the negative values of slopes of the $A = f(t)$ dependencies could be treated as the rate constants of MB photodegradation (Table 6).

Table 6. Rate constants of MB degradation for sole MB, pure PMMA, and PMMA with addition of the studied materials.

Sample	10^3 Rate Constant, h^{-1}	Sample	10^3 Rate Constant, h^{-1}	10^3 Rate Constant in Reference to PMMA, h^{-1}
sole MB in darkness	0.35 ± 0.15	(**1**) in light	4.03 ± 0.32	1.51 ± 0.36
sole MB in light	1.52 ± 0.09	(**2**) in light	2.95 ± 0.15	0.43 ± 0.22
PMMA in darkness	0.26 ± 0.04	(**3**) in light	2.53 ± 0.12	0.01 ± 0.20
PMMA in light	2.52 ± 0.16	(**4**) in light	3.19 ± 0.22	0.67 ± 0.27

Table 6 shows that the rate of MB degradation in darkness in case of sole MB and pure PMMA was very low and were similar. Under UV light, MB degraded more than 4 times faster. The PMMA foil accelerated its degradation by an additional 1.5 times (from 1.52×10^{-3} h^{-1} to 2.52×10^{-3} h^{-1}). A correct comparison of the rate constants for PMMA foil with addition of titanium complexes should be referenced to the rate constant for pure PMMA foil "in darkness" (the last column in Table 6). The values of the corrected rate constants, presented in Table 6, indicated a trace photocatalytic activity of (**3**) in the MB degradation process. Complexes (**2**) and (**4**) were slightly more active and the para derivative was the best among those three complexes. The complex with fluorene (**1**) was clearly the most active in studied conditions.

3.7. ESR Evidence of Paramagnetic Species

To investigate the mechanism of photocatalytic process, the ESR analysis of powdered compounds and prepared materials was performed. Three types of paramagnetic species were observed for studied compounds (Table 7). Spectra of sample (**1**) (powder and PMMA foil) exhibited very weak radical signal (g_{exp} = 2.0031 and 2.0036, respectively) before UV-Vis irradiation. Due to unresolved g-tensor components, it was difficult to state with certainty what the structure of this radical was, but the value of the parameter indicated that it was O_2^- rather than O^- [43]. Surprisingly, compound (**4**), before irradiation, exhibited signals of both types of oxygen radical (in case of powdered sample) and strong signal of O^- in case of the PMMA foil. For the other two samples, paramagnetic species before irradiation were not found. The most interesting ESR spectrum was observed for irradiated foil with sample (**1**) (with fluorene) (Figure 7). Spectrum—represented by signals originated in Ti(III) (centered in a distorted octahedral coordination field [44]), O_2^-, and O^- parameters—are given in Table 7. Similar spectrum was observed for the same but powdered sample, although the signals were much weaker, and Ti(III) signal was observed only as a trace. Irradiated sample (**2**) (foil and powder) exhibited spectra for which simulated g-parameter tensor components are typical of orthorhombic superoxide diatomic oxygen O_2^- adsorbed on metal oxide surfaces [45]. However, the spectrum of foil was much better resolved than the spectrum of powder; signals observed for both kinds of samples and were characterized by the lowest intensity within the investigated samples. The signal is the result of electron trapping on Ti(IV) surface [46]:

$$e^- + Ti^{4+}OH \longrightarrow Ti^{3+}OH \tag{2}$$

$$Ti^{3+}OH + O_2 \longrightarrow Ti^{4+}OH + \bullet O_2^- \tag{3}$$

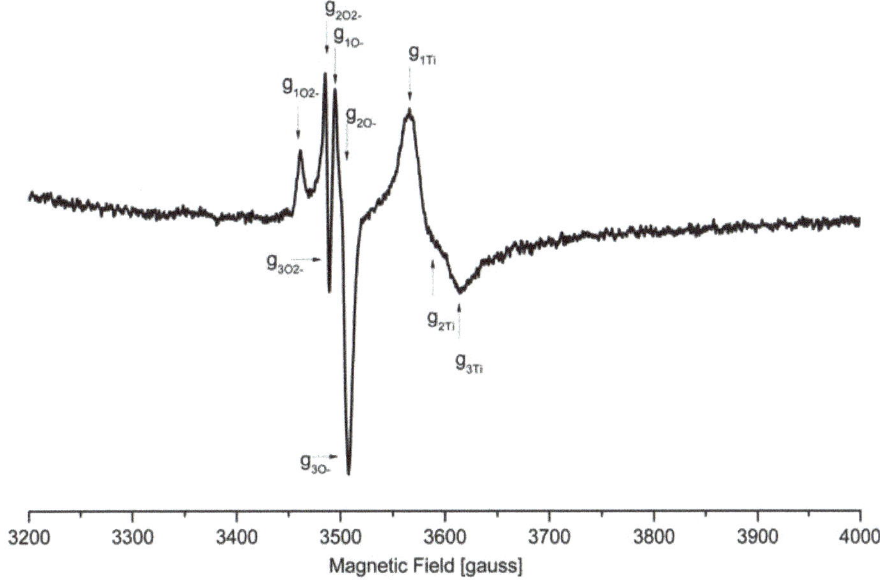

Figure 7. EPR spectrum of UV-Vis irradiated PMMA foil with complex (**1**).

Table 7. Results of EPR studies for UV-Vis irradiated complexes (PMMA foils) (($Ti_4O_2(O^iBu)_{10}(O_2CR')_2$) ($R' = C_{13}H_9$ (**1**), PhCl (**2**), PhNO$_2$ (**3**), PhNH$_2$ (**4**)).

Sample	g Parameters			Species
PMMA +	g_1	g_2	g_3	
	2.024	2.0095	2.0034	O_2^-
(**1**)	1.967	1.957	1.938	Ti(III)
	2.003	-	1.997	O^-
(**2**)	2.024	2.0095	2.0034	O_2^-
(**3**)	2.0185	2.0052	1.987	O^-
(**4**)	2.0182	2.005	1.987	O^-

The main paramagnetic species observed on the spectra of samples (**3**) and (**4**) was V-type hole center [47]. Creation of this paramagnetic O^- species could be the result of the electron removal and radical being stabilized by reduction of Ti(IV) to Ti(III) [40]; the signal of Ti(III) might have been not observed due its broadness and nonintense line.

4. Discussion

The reaction of Ti(O^{i}Bu)$_4$ with fluorene-9-carboxylate acid, 3-chlorobenzoic acid, and 3-nitrobenzoic acid in molar ratio 4:1 at room temperature led to the formation of ($Ti_4O_2(O^iBu)_{10}(O_2CR')_2$) ($R' = C_{13}H_9$ (**1**), C_6H_4Cl (**2**), $C_6H_4NO_2$ (**3**)) oxo-clusters. Structural characterization of these compounds (single crystals X-ray diffraction and vibrational spectroscopy) and their comparison with the earlier studied ($Ti_4O_2(O^iBu)_{10}(O_2CC_6H_4NH_2)_2$) complex (**4**) [27] did not show a significant impact of the carboxylate ligand type on the {Ti$_4$(μ_4–O)(μ–O)} core structure. However, the clear {Ti$_2$(μ_2–O)} angle changes (which could be responsible for the facilitation of the photocatalytic process [26]) were observed, i.e., (**4**) < (**1**) < (**2**) < (**3**) (Table 2). The substantial difference in band structures based on used ligands revealed the results of DFT calculations. The presence of such carboxylic ligands as $-O_2CC_{13}H_9$, $-O_2CC_6H_4Cl$, and $-O_2CC_6H_4NH_2$ in structures of (**1**), (**2**), and (**4**), respectively, resulted in the *n*-doped band structure, while $-O_2CC_6H_4NO_2$ (**3**) was *p*-doped. The comparison of photoluminescence spectra of pure carboxylic ligands and synthesized compounds indicated the distinctive interactions between central oxo-titanium core and external ligands. Those interactions influenced the band gap value of produced materials, i.e., 3.59 eV for (**2**), 2.98 eV (**3**), 2.57 eV (**4**), and 2.55 eV (**1**) (the band gap of the widely used TiO$_2$ anatase is 3.20 eV [22]). The obtained results were comparable with the previously studied Ti(IV) oxo-clusters containing functionalized carboxylate ligands [20]. The low value of the band gap of (**1**) can be explained by the presence of fluorene groups, whose derivatives are known for their photoluminescence properties [48].

Isolated crystals of all synthesized Ti(IV) oxo-complexes exhibited hydrophobic properties and very low sensitivity to the moisture. Therefore, in order to estimate the photocatalytic activity of the obtained compounds in the MB degradation process (UV irradiated), the composites produced by the introduction of (**1**)–(**4**) crystalline powders into the polymer matrix were used. The choice of PMMA as a polymer matrix was made on the basis of the spectroscopic criterion, i.e., the lack of PMMA bands in the Raman spectrum ranges in which the characteristic bands of oxo-complexes appear. This enabled the confirmation of the oxo-complex presence in the polymer matrix and the evaluation of potential interactions between the complex particles and the polymer matrix. Analysis of SEM images showed that PMMA + (**3**) foil contained dispersed large grains of complex (**3**) (Figure 5f), which was confirmed by the Raman spectrum of this composite (low intense bands at 1614 cm^{-1}, 1659 cm^{-1} (ν(COO)), 1534 cm^{-1} (ν(NO$_2$), 1008 cm^{-1} (ν(C–O), and 697,653 cm^{-1} (ν(Ti–O–Ti), (Figure 4, Table 3). Simultaneously, the lack of significant differences in the position and the shape of PMMA bands in PMMA + (**3**) spectrum suggested that significant interactions between polymer matrix and dispersed grains of (**3**) were not formed (Figure 4). The MB photodegradation studies (in UV) exhibited the trace activity of the PMMA + (**3**) composite, which could be associated with the *p*-doped band structure of

(**3**) (other synthesized oxo-complexes are *n*-doped), the high value of band gap (2.98 eV). It is probable that the dispersion of large grains (micrograins/microcrystals) of the (**3**) complex in the PMMA matrix also contributed to the poor activity of this composite. The results obtained for (**3**) were in good agreement with earlier investigations of Liu et al. on the 3-nitrocarboxylate substituted compound $(Ti_6O_4(O^iPr)_{10}(O_3P\text{-}Phen)_2(O_2CC_6H_4NO_2)_2)$ that also did not show any activity towards photocatalytic hydrogen production [23]. A better photocatalytic activity was observed for the PMMA + (**2**) foil (Figure 6 and Table 6). In addition, the spectral data analysis in that case did not indicate the formation interactions between (**2**) and PMMA (Figure 4, Table 3). Considering earlier reports [21], it should be noted that the presence of $-O_2CC_6H_4Cl$ ligands in the structure of (**2**) resulted in an increase in the value of the band gap up to 3.59 eV, which might have contributed to the lower photocatalytic activity of this material. Analysis of SEM images of PMMA + (**2**) showed that nanoparticles of (**2**)—whose diameter (*d*) is lower than 25 nm—and grains of *d* = 50 nm–300 μm formed a filling of a polymer matrix (Figure 5c,d). The substitution of the $\{Ti_4O_2\}$ core by $-O_2CC_6H_4NH_2$ ligands promoted the decrease of band gap value (2.57 eV [20,26]) which was noted for the complex (**4**). Simultaneously, the SEM studies of PMMA + (**4**) foil proved that the filling of the polymer matrix was formed by islands composed of nanoparticles of *d* = 30–60 nm (Figure 5h). Distinct differences in the intensity and width of bands in Raman spectra of both oxo-cluster (**4**) and PMMA (Figure 4) suggested the formation of hydrogen bonds between the particles of the complex and the polymer (the hydrogen bonds between $-NH_2$ groups of oxo-clusters and $-C=O$ groups of the polymer are possible). In this case, low value of the band gap and possible weak interactions between amine groups of the oxo-cluster and PMMA had a direct influence on the increase of the photocatalytic activity of complex (**4**). A noticeable high photocatalytic activity was observed for foils produced by introduction of complex (**1**) into the PMMA matrix. Probable substitution of $\{Ti_4O_2\}$ cores by fluorene-9-carboxylate ligands impacted the decrease of the band gap value (2.88 eV). Analysis of SEM images revealed that the densely packed nanoparticles of diameters *d* = 20–40 nm were the main filling of the PMMA + (**1**) foil (Figure 5b). The bands that were found in the Raman spectrum of PMMA + (**1**) at 1611 (m), 1581 (w) cm^{-1}, 1023 (w) cm^{-1}, and 660–700 (vw) cm^{-1} confirmed the presence of (**1**) in the polymer matrix (Figure 4, Table 3). Moreover, the comparison of PMMA + (**1**) and PMMA Raman spectra indicated lack of interactions between complex particles and the polymer matrix. It is interesting that ESR studies revealed the appearance of Ti(III) states during its irradiation from only the PMMA + (**1**) foil among the produced composite materials. According to works of Snoeberger et al. [49] and Negre et al. [50], the type of doping influences the mechanism of photoinduced charge transfer, promoting either electron or hole injection into functionalized oxo-titanium cluster [49,50]. Considering the result of the ESR and DFT studies, we can conclude that *n*-doping of POTs (in the case of (**3**)) did not allow for the photocatalytic activity due to absence of photogenerated Ti(III) states upon irradiation, while *p*-doping made it possible to generate Ti(III) and induce the photocatalytic action. DFT results revealed that HOMO orbitals of (**1**), (**2**), and (**4**) were located on ligands and LUMO on Ti atoms of the oxo-titanium core. Thus, upon irradiation, the charge was transferred to the core, generating Ti(III) states. This was confirmed by ESR experiment, which revealed Ti(III) states in PMMA–(**1**) composite foiled upon irradiation. According to the theoretical approach, these states should be present in (**2**) and (**4**), but the experiment did not confirm it. Possible explanation involves slower recombination of charges in compound with fluorene (**1**) than in ones with benzoic acid derivatives (**2**) and (**4**), which allows the detection of Ti(III).

5. Summary

The conducted works have led to an isolate and structural characterization of three new tetranuclear Ti(IV) oxo-clusters (TOCs) of the general formula $(Ti_4O_2(O^iBu)_{10}(OOCR')_2)$ from the 4:1 reaction mixture of $Ti(O^iBu)_4$ with organic acids such as fluorene-9-carboxylate acid (**1**), 3-chlorobenzoic acid (**2**), and 3-nitrobenzoic acid (**3**). Analysis of structural and spectroscopic data exhibited that substitution of the $\{Ti_4O_2\}$ core by the different carboxylate ligands did not significantly affect its architecture (the clear changes were found only for the value of Ti–(μ–O)–Ti angles). However, the use of different carboxylate ligands allowed control of the band gap value of produced oxo-clusters in the range of 3.6–2.6 eV. The photocatalytic activity of (**1**)–(**4**) was estimated on the basis of MB photodegradation processes (UV irradiated) that proceeded on the surface of a composite foils produced by the introduction of TOCs (20 wt %) into the PMMA matrix (PMMA + TOCs). The trace photocatalytic activity was noticed for (**3**), i.e., the *p*-doped $(Ti_4O_2(O^iBu)_{10}(O_2CC_6H_4NO_2)_2)$ complex ((**1**), (**2**), (**4**) were *n*-doped). The highest photocatalytic activity was evidenced for PMMA + (**1**) foil, which contained $(Ti_4O_2(O^iBu)_{10}(O_2CC_{13}H_9)_2)$ (**1**) complex. According to results of our works, the following factors influence photoactivity of (**3**): *n*-dope character of the compound, relatively low band gap (2.55 eV), the presence of Ti(III) states upon irradiation evidenced by ESR spectrometry, and equal nanocrystals distribution in the PMMA matrix (confirmed by SEM imaging).

Supplementary Materials: The following are available online at http://www.mdpi.com/1996-1944/11/9/1661/s1, Figure S1: Infrared and Raman spectra of (**1**)–(**3**) complexes, Figure S2: The calculated partial density of states of oxo-complexes (**1**), (**2**), and (**3**), Figure S3: DFT calculated HOMO (top) and LUMO (bottom) molecular orbitals of (**1**), (**2**) and (**3**), **Table S1:** Comparison of the experimental and the calculated (DFT) structural data of $\{Ti_4O_2\}$ cores of studied oxo-clusters: $[Ti_4O_2(O^iBu)_{10}(O_2CC_{13}H_9)_2]$ (**1**), $[Ti_4O_2(O^iBu)_{10}(O_2CC_6H_4Cl)_2]$ (**2**), $[Ti_4O_2(O^iBu)_{10}(O_2CC_6H_4NO_2)_2]$ (**3**), $[Ti_4O_2(OMe)_{10}(O_2CC_{13}H_9)_2]$ DFT(**1**), $[Ti_4O_2(OMe)_{10}(O_2CC_6H_4Cl)_2]$ DFT(**2**), $[Ti_4O_2(OMe)_{10}(O_2CC_6H_4NO_2)_2]$ DFT (**3**).

Author Contributions: Conceptualization, Piotr Piszczek, Maciej Janek, Methodology, conceived, and designed of experiments; Maciej Janek; Investigations and Formal Analysis; Maciej Janek, Tadeusz M. Muzioł, Aleksandra Radtke, Maria Jerzykiewicz, Piotr Piszczek; Writing-Review & Editing Piotr Piszczek, Maciej Janek; Supervision, Piotr Piszczek.

Acknowledgments: Diffraction data have been collected on BL14.X (X = (1, 2 ,3)) at the BESSY II electron storage ring operated by the Helmholtz-Zentrum Berlin [51]. We would particularly like to acknowledge the help and support of Piotr Małecki during the experiment.

Conflicts of Interest: The authors declare no conflict of interest.

References

1. Chen, X.; Mao, S.S. Titanium dioxide nanomaterials: Synthesis, Properties, Modifications, and Applications. *Chem. Rev.* **2007**, *107*, 2891–2959. [CrossRef] [PubMed]

2. Hashimoto, K.; Irie, H.; Fujishima, A. TiO_2 Photocatalysis: A Historical Overview and Future Prospects. *Jpn. J. Appl. Phys.* **2005**, *44*, 8269–8285. [CrossRef]

3. Etacheri, V.; Di Valentin, C.; Schneider, J.; Bahnemann, D.; Pillai, S.C. Visible-light activation of TiO_2 photocatalysts: Advances in theory and experiments. *J. Photochem. Photobiol. C* **2015**, *25*, 1–29. [CrossRef]

4. Carp, O.; Huisman, C.L.; Reller, A. Photoinduced reactivity of titanium dioxide. *Prog. Solid State Chem.* **2004**, *32*, 33–177. [CrossRef]

5. Fujishima, A.; Zhang, X.; Tryk, D.A. TiO_2 photocatalysis and related surface phenomena. *Surf. Sci. Rep.* **2008**, *63*, 515–582. [CrossRef]

6. Li, N.; Matthews, P.D.; Luo, H.-K.; Wright, D.S. Novel properties and potential applications of functional ligand-modified polyoxotitanate cages. *Chem. Commun.* **2016**, *52*, 11180–11190. [CrossRef] [PubMed]

7. Benedict, J.B.; Freindorf, R.; Trzop, E.; Cogswell, J.; Coppens, P. Large Polyoxotitanate Clusters: Well-Defined Models for Pure-Phase TiO2 Structures and Surfaces. *J. Am. Chem. Soc.* **2010**, *132*, 13669–13671. [CrossRef] [PubMed]

8. Benedict, J.B.; Coppens, P. The Crystalline Nanocluster Phase as a Medium for Structural and Spectroscopic Studies of Light Absorption of Photosensitizer Dyes on Semiconductor Surfaces. *J. Am. Chem. Soc.* **2010**, *132*, 2938–2944. [CrossRef] [PubMed]

9. Coppens, P.; Chen, Y.; Trzop, E. Crystallography and Properties of Polyoxotitanate Nanoclusters. *Chem. Rev.* **2014**, *114*, 9645–9661. [CrossRef] [PubMed]

10. Ohlmaier-Delgadillo, F.; Castillo-Ortega, M.M.; Ramirez-Bon, R.; Armenta-Villegas, L.; Rodriguez-Félix, D.E.; Santacruz-Ortega, H.; del Castillo-Castro, T.; Santos-Sauceda, I. Photocatalytic properties of PMMA-TiO_2 class I and class II hybrid nanofibers obtained by electrospinning. *J. Appl. Polym. Sci.* **2016**, *133*, 44334. [CrossRef]

11. Kickelbick, G. *Hybrid Materials: Synthesis, Characterization, and Applications*; Wiley-VCH Verlag GmbH & Co. KGaA: Weinheim, Germany, 2007.

12. Schubert, U. Cluster-Based Inorganic-Organic Hybrid Materials. *Chem. Soc. Rev.* **2011**, *40*, 575–582. [CrossRef] [PubMed]

13. Gross, S. Oxocluster-Reinforced Organic-Inorganic Hybrid Materials: Effect of Transition Metal Oxoclusters on Structural and Functional Properties. *J. Mater. Chem.* **2011**, *21*, 15853–15861. [CrossRef]

14. Carraro, M.; Gross, S. Hybrid Materials Based on the Embedding of Organically Modified Transition Metal Oxoclusters or Polyoxometalates into Polymers for Functional Applications: A Review. *Materials* **2014**, *7*, 3956–3989. [CrossRef] [PubMed]

15. Narayanam, N.; Chintakrinda, K.; Fang, W.-H.; Kang, Y.; Zhang, L.; Zhang, J. Azole Functionalized Polyoxo-Titanium Clusters with Sunlight-Driven Dye Degradation Applications: Synthesis, Structure, and Photocatalytic Studies. *Inorg. Chem.* **2016**, *55*, 10294–10301. [CrossRef] [PubMed]

16. Horiuchi, Y.; Toya, T.; Saito, M.; Mochizuki, K.; Iwata, M.; Higashimura, H.; Anpo, M.; Matsuoka, M. Visible-Light-Promoted Photocatalytic Hydrogen Production by Using an Amino-Functionalized Ti(IV) Metal-Organic Framework. *J. Phys. Chem. C* **2012**, *116*, 20848–20853. [CrossRef]

17. Rozes, L.; Sanchez, C. Titanium oxo-clusters: Precursors for a Lego-like construction of nanostructured hybrid materials. *Chem. Soc. Rev.* **2011**, *40*, 1006–1030. [CrossRef] [PubMed]

18. Piszczek, P.; Radtke, A.; Muzioł, T.; Richert, M.; Chojnacki, J. The conversion of multinuclear μ-oxo titanium(IV) species in the reaction of Ti(O^{i}Bu)$_4$ with branched organic acids; results of structural and spectroscopic studies. *Dalton Trans.* **2012**, *41*, 8261–8269. [CrossRef] [PubMed]

19. Radtke, A.; Piszczek, P.; Muzioł, T.; Wojtczak, A. The Structural Conversion of Multinuclear Titanium(IV) μ-Oxo-complexes. *Inorg. Chem.* **2014**, *53*, 10803–10810. [CrossRef] [PubMed]

20. Fang, W.-H.; Zhang, L.; Zhang, J. Synthetic strategies, diverse structures and tunable properties of polyoxo-titanium clusters. *Chem. Soc. Rev.* **2018**, *47*, 404–421. [CrossRef] [PubMed]

21. Schubert, U. Chemical modification of titanium alkoxides for sol-gel processing. *J. Mater. Chem.* **2005**, *15*, 3701–3715. [CrossRef]

22. Liu, J.-X.; Gao, M.-Y.; Fang, W.-H.; Zhang, L.; Zhang, J. Bandgap Engineering of Titanium–Oxo Clusters: Labile Surface Sites Used for Ligand Substitution and Metal Incorporation. *Angew. Chem. Int. Ed.* **2016**, *55*, 5160–5165. [CrossRef] [PubMed]

23. Kim, S.; Sarkar, D.; Kim, Y.; Park, M.H.; Yoon, M.; Kim, Y.; Kim, M. Synthesis of functionalized titanium-carboxylate molecular clusters and their catalytic activity. *J. Ind. Eng. Chem.* **2017**, *53*, 171–176. [CrossRef]

24. Lin, Y.; Zhou, Y.-F.; Chen, Z.-H.; Lui, F.-H.; Zhao, L.; Su, Z.-M. Synthesis, structure, and photocatalytic hydrogen of three environmentally friendly titanium oxo-clusters. *Inorg. Chem. Commun.* **2014**, *40*, 22–25. [CrossRef]

25. Wu, Y.-Y.; Luo, W.; Wang, Y.-H.; Pu, Y.-Y.; Zhang, X.; You, L.-S.; Zhu, Q.-Y.; Dai, J. Titanium–oxo–Clusters with Dicarboxylates: Single-Crystal Structure and Photochromic Effect. *Inorg. Chem.* **2012**, *51*, 8982–8988. [CrossRef] [PubMed]

26. Luo, W.; Ge, G. Two Titanium-oxo-Clusters with Malonate and Succinate Ligands: Single-Crystal Structures and Catalytic Property. *J. Clust. Chem.* **2016**, *27*, 635–643. [CrossRef]

27. Janek, M.; Muzioł, T.; Piszczek, P. The structure and photocatalytic activity of the tetranuclear titanium(IV) oxo-complex with 4-aminobenzoate ligands. *Polyhedron* **2018**, *141*, 110–1178. [CrossRef]

28. Liu, Z.; Lei, J.; Frasconi, M.; Li, X.; Cao, D.; Zhu, Z.; Schneebeli, S.T.; Schatz, G.C.; Stoddart, J.F. A Square-Planar Tetracoordinate Oxygen-Containing Ti$_4$O$_{17}$ Cluster Stabilized by Two 1,1'-Ferrocenedicarboxylato Ligands. *Angew. Chem. Int. Ed.* **2014**, *51*, 9193–9197. [CrossRef] [PubMed]

29. *CrysAlis RED and CrysAlis CCD*; Oxford Diffraction Ltd.: Abingdon, Oxfordshire, England, 2000.

30. Krug, M.; Weiss, M.S.; Heinemann, U.; Mueller, U. XDSAPP: A graphical user interface for the convenient processing of diffraction data using XDS. *J. Appl. Crystallogr.* **2012**, *45*, 568–572. [CrossRef]

31. Kabsch, W. XDS. *Acta Cryst. D* **2010**, *66*, 125–132. [CrossRef] [PubMed]

32. Sheldrick, G.M. Crystal structure refinement with SHELXL. *Acta Cryst. C* **2015**, *71*, 3–8. [CrossRef] [PubMed]

33. Brandenburg, K. *Diamond*; Release 2.1e; Crystal Impact GbR: Bonn, Germany, 2001.

34. Farrugia, L.J. WinGX and ORTEP for Windows: An update. *J. Appl. Crystallogr.* **2012**, *45*, 849–854. [CrossRef]

35. Frisch, M.J.; Trucks, G.W.; Schlegel, H.B.; Scuseria, G.E.; Robb, M.A.; Cheeseman, J.R.; Scalman, G.; Barone, V.; Mennucci, B.; Petersson, G.A.; et al. *Gaussian09*; Revision D.01; Gaussian, Inc.: Wallingford, CT, USA, 2013.

36. O'Boyle, N.M.; Tenderholt, A.L.; Langner, K.M. Cclib: A library for package-independent computational chemistry algorithms. *J. Comp. Chem.* **2008**, *29*, 839–845. [CrossRef] [PubMed]

37. Boyle, T.J.; Alam, T.M.; Tafoya, C.J.; Scott, B.L. Formic Acid Modified $Ti(OCHMe_2)_4$. Syntheses, Characterization, and X-ray Structures of $Ti_4(\mu_4\text{-}O)(\mu\text{-}O)(OFc)_2(\mu\text{-}OR)_4(OR)_6$ and $Ti_6(\mu_3\text{-}O)_6(OFc)_6(OR)_6$ ($OFc = O_2CH$; $OR = OCHMe_2$). *Inorg. Chem.* **1998**, *37*, 5588–5594. [CrossRef] [PubMed]

38. Eslava, S.; Hengesbach, F.; McPartlin, M.; Wright, D.S. Heterometallic cobalt(II)–titanium(IV) oxo cages; key building blocks for hybrid materials. *Chem. Commun.* **2010**, *46*, 4701–4703. [CrossRef] [PubMed]

39. Eslava, S.; McPartlin, M.; Thomson, R.I.; Rawson, J.M.; Wright, D.S. Single-Source Materials for Metal-Doped Titanium Oxide: Syntheses, Structures, and Properties of a Series of Heterometallic Transition-Metal Titanium Oxo Cages. *Inorg. Chem.* **2010**, *49*, 11532–11540. [CrossRef] [PubMed]

40. Wendlandt, W.W.; Hecht, H.G. *Reflectance Spectroscopy*; Interscience Publishers: New York, NY, USA, 1966.

41. Nowak, M.; Kauch, B.; Szperlich, P. Determination of energy band gap of nanocrystalline SbSI using diffuse reflectance spectroscopy. *Rev. Sci. Instrum.* **2009**, *80*, 046107. [CrossRef] [PubMed]

42. Garza, A.J.; Scuseria, G.E. Predicting Band Gaps with Hybrid Density Functionals. *J. Phys. Chem. Lett.* **2016**, *7*, 4165–4170. [CrossRef] [PubMed]

43. Suriye, K.; Lobo-Lapidus, R.J.; Yeagle, G.J.; Praserthdam, P.; Britt, D.; Gates, B.C. Probing Defect Sites on TiO_2 with $[Re_3(CO)_{12}H_3]$: Spectroscopic Characterization of the Surface Species. *Chem. Eur. J.* **2008**, *14*, 1402–1414. [CrossRef] [PubMed]

44. Wu, Y.Y.; Lu, X.-W.; Qi, M.; Su, H.C.; Zhao, X.-W.; Zhu, Q.-Y.; Dai, J. Titanium–Oxo Cluster with 9-Anthracenecarboxylate Antennae: A Fluorescent and Photocurrent Transfer Material. *Inorg. Chem.* **2014**, *53*, 7233–7240. [CrossRef] [PubMed]

45. Dan-Hardi, M.; Serre, C.; Frot, T.; Rozes, L.; Maurin, G.; Sanchez, C.; Fe´rey, G. A New Photoactive Crystalline Highly Porous Titanium(IV) Dicarboxylate. *J. Am. Chem. Soc.* **2009**, *131*, 10857–10859. [CrossRef] [PubMed]

46. Xiong, L.B.; Li, J.L.; Yang, B.; Yu, Y. Ti^{3+} in the Surface of Titanium Dioxide: Generation, Properties and Photocatalytic Application. *J. Nanomater.* **2012**, *2012*, 9. [CrossRef]

47. Prakash, A.M.; Kurshev, K.L. Electron Spin Resonance and Electron Spin Echo Modulation Evidence for the Isomorphous Substitution of Ti in TAPO-5 Molecular Sieve. *J. Phys. Chem. B* **1997**, *101*, 9794–9799. [CrossRef]

48. Agarwal, N.; Nayak, P.K.; Periasamy, N. Synthesis, photoluminescence and electrochemical properties of 2,7-diarylfluorene derivatives. *J. Chem. Sci.* **2008**, *120*, 355–362. [CrossRef]

49. Soeberger III, R.C.; Young, K.J.; Tang, J.; Allen, L.J.; Crabtree, R.H.; Brudvig, G.W.; Coppens, P.; Batista, V.S.; Benedict, J.B. Interfacial Electron Transfer into Functionalized Crystalline Polyoxotitanate Nanoclusters. *J. Am. Chem. Soc.* **2012**, *134*, 8911–8917. [CrossRef] [PubMed]

50. Negre, C.F.A.; Young, K.J.; Oviedo, M.B.; Allen, L.J.; Sánchez, C.G.; Jarzembska, K.N.; Benedict, J.B.; Crabtree, R.H.; Coppens, P.; Brudvig, G.W.; et al. Photoelectrochemical Hole Injection Revealed in Polyoxotitanate Nanocrystals Functionalized with Organic Adsorbates. *J. Am. Chem. Soc.* **2014**, *136*, 16420–16429. [CrossRef] [PubMed]

51. Mueller, U.; Förster, R.; Hellmig, M.; Huschmann, F.U.; Kastner, A.; Malecki, P.; Pühringer, S.; Röwer, M.; Sparta, K.; Steffien, M.; et al. The macromolecular crystallography beamlines at BESSY II of the Helmholtz-Zentrum Berlin: Current status and perspectives. *Eur. Phys. J. Plus* **2015**, *130*, 141–150. [CrossRef]

© 2018 by the authors. Licensee MDPI, Basel, Switzerland. This article is an open access article distributed under the terms and conditions of the Creative Commons Attribution (CC BY) license (http://creativecommons.org/licenses/by/4.0/).

 *materials*

Article

Stone/Coating Interaction and Durability of Si-Based Photocatalytic Nanocomposites Applied to Porous Lithotypes

Marco Roveri [1,*], Francesca Gherardi [2], Luigi Brambilla [1], Chiara Castiglioni [1] and Lucia Toniolo [1]

[1] Politecnico di Milano, Dipartimento di Chimica, Materiali e Ingegneria Chimica "G. Natta", 20133 Milano, Italy; luigi.brambilla@polimi.it (L.B.); chiara.castiglioni@polimi.it (C.C.); lucia.toniolo@polimi.it (L.T.)
[2] School of Chemistry, University of Lincoln, LN6 7DL Lincoln, UK; fgherardi@lincoln.ac.uk
* Correspondence: marco.roveri@polimi.it; Tel.: +39-02-2399-3143

Received: 6 October 2018; Accepted: 12 November 2018; Published: 15 November 2018

Abstract: The use of hybrid nanocoatings for the protection of natural stones has received increasing attention over the last years. However, the interaction of these materials with stones and, in particular, its modification resulting from the blending of nanoparticles and matrices, are yet little explored. In this work, the interaction of two nanocomposite coatings (based on alkylalkoxysilane matrices and TiO_2 nanoparticles in water and 2-propanol) with two different porous stones is examined in detail by comparing their absorption behaviour and protection performance with those of the respective TiO_2-free matrices. It is shown that the protective effectiveness of both matrices is not negatively affected by the presence of TiO_2, as the desired water barrier effect is retained, while a significant photocatalytic activity is achieved. The addition of titania leads to a partial aggregation of the water-based matrix and accordingly reduces the product penetration into stones. On the positive side, a chemical interaction between titania and this matrix is observed, probably resulting in a greater stability of nanoparticles inside the protective coating. Moreover, although an effect of TiO_2 on the chemical stability of matrices is observed upon UV light exposure, the protective performance of coatings is substantially maintained after ageing, while the interaction between matrices and nanoparticles results in a good retention of the latter upon in-lab simulated rain wash-out.

Keywords: TiO_2 nanoparticles; alkylalkoxysilane; stone protection; water-repellency; photocatalysis; UV ageing; artificial rain; photo-oxidative degradation; durability

---

1. Introduction

The use of TiO_2 nanoparticles as protective treatments for natural stones has aroused much interest over the last 10 years [1–5], prompting efforts to enhance their effectiveness as photocatalytic agents in this specific field of application. Indeed, one of the main drawbacks of using nanoTiO_2 dispersions is that they have poor adhesion to the stone substrates, thus tending to penetrate into the pore structure or be easily removed by the mechanical action of wind and rainfall. In both cases, this results in a significant decrease in the titania surface content and thus in lower photocatalytic performances [6,7]. One general strategy towards overcoming this issue has been represented by the development of TiO_2-based nanocomposites, which make up a wide class of hybrid materials obtained from the addition of titania nanoparticles to organo-modified silica precursors such as TEOS/alkylalkoxysilanes or different siloxane/acrylic/fluorinated/epoxy polymer dispersions [8–22]. These composite materials have displayed a number of advantages over bare nanoTiO_2 dispersions, which include a better adhesion to the stone substrates, a lower tendency towards unwanted nanoparticle aggregation and the ability to combine a self-cleaning photocatalytic action with the properties of traditional water-repellent or consolidation treatments [20–22], while retaining a good aesthetic compatibility. Indeed, a strict requirement in this kind of application is that coatings should not change the "appearance" of stones, that is, their surface colour, texture and finishing.

From the viewpoint of materials characterization, the properties of some of these hybrid coatings have been very well described: in particular, investigations have focused on elucidating the structure of the titania-polymer interface, pointing out the stabilizing effect of filler-matrix interactions and the modification of the wetting behaviour of the composites due to the rough surface topography induced by TiO_2 nanoparticles [23,24]. Furthermore, as regards the application on stone, the protective performance of several TiO_2-based nanocomposites has been assessed on different stone substrates, including both high and low porosity stones [11,13,22]. Specifically, the contribution of nanoparticles to the reduction of surface wettability has been shown on low porosity stones such as marbles [19,21], while the photocatalytic performance and self-cleaning behaviour of $nanoTiO_2$ and the influence of different polymer matrices have been widely investigated on both high and low porosity stones [11,19,22].

However, few studies did investigate in more detail how the combination of nanoparticles and polymer matrices affects the properties of the coatings in their interaction with stone substrates [10,11,19]. Furthermore, there is still very limited information on the durability of these coatings under real outdoor conditions [21,25,26]. Accelerated ageing procedures have indeed been performed to assess the effects of rain wash-out [27–30], wet-dry cycles [31], exposure to UV light [26,27,29–32] and soiling [27,30], especially with regard to the adhesion of titania nanoparticles to the stone substrates and the evaluation of the change in their photocatalytic activity. However, the results obtained on different stones and cementitious materials are far from univocal and it has been suggested that the intrinsic properties of stones play a relevant role in the long-term efficiency of the products and result in different durability issues [25,26,33]. Furthermore, only few studies have set out to address the influence of photo-active TiO_2 nanoparticles on the durability of hybrid coatings upon solar light irradiation [20,34] and, in particular, to clarify the effects on their hydrophobic properties due to the inherent UV-induced superhydrophilicity of titania and to the possible photo-catalysed oxidative degradation of matrices.

In this work, which is part of a wider-scope research dealing with the set-up and testing of innovative nanocomposite materials for the conservation of architectural heritage, the study of the interaction of two different TiO_2-based nanocomposite formulations with two porous stones of high relevance in the built heritage field is addressed. These formulations, consisting of alkylalkoxysilane reactive sols combined with different titania nanoparticles, were developed in the framework of the EU-funded NanoCathedral project [35] and their performance has already been discussed in a previous study [36]. In the present research, the interaction of these nanocomposites with stones and their protective performance are compared to those of the corresponding alkylalkoxysilane matrices, especially in terms of absorption behaviour, surface textural modification and achieved water-repellency, focusing on the role played therein by the combination of TiO_2 nanoparticles with matrices. Then, an investigation of the durability of nanocomposites and matrices is carried out in order to assess the contribution of TiO_2 to the possible photo-oxidative degradation of nanocomposites upon UV light exposure and check the retention of nanoparticles on the stone surfaces upon rain wash-out.

2. Materials and Methods

Two different high porosity stones (Figure 1), characterized by a calcareous or siliceous composition, were used. Ajarte (Lumaquela de Ajarte) is a sedimentary rock from Treviño area (Spain), having a calcite matrix with a high amount of intercrystalline pores. Obernkirchen is a quartzarenite from Bückeberge area (Germany), characterized by a fine-grained texture. These natural stones have particular relevance in the cultural heritage field: Ajarte limestone was widely employed as building material in the north of Spain since the Middle Ages and is found in monuments such as

St. Mary's Cathedral in Vitoria-Gasteiz (XIII-XVI cent.), while Obernkirchen sandstone is especially renowned for being one of the materials used in the construction of Cologne Cathedral (XIII-XIX cent.).

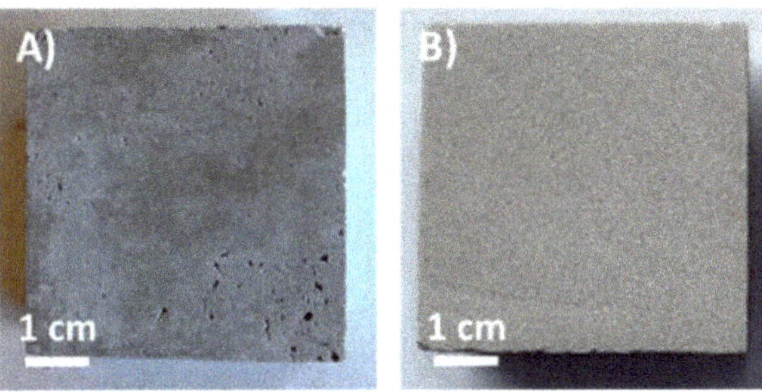

Figure 1. Photographs of Ajarte (**A**) and Obernkirchen (**B**) stone specimens.

The two nanocomposite formulations used in this study, hereafter referred to as WNC and ANC, were developed in the framework of the EU-funded H2020 NanoCathedral project (Grant Agreement n. 646178) by two SME Partners. They consist of TiO_2 nanoparticles (Colorobbia Consulting srl, Sovigliana Vinci, Italy) dispersed in commercial alkylalkoxysilane reactive sols (ChemSpec srl, Peschiera Borromeo, Italy). The details of the preparation, including the commercial names of product components, are protected by a non-disclosure agreement. The main properties of these components are shown in Table 1, while the composition of the formulates is reported in Table 2.

Table 1. Main properties of the components used for the preparation of WNC and ANC, namely TiO_2 NPs ($nTiO_2$-W and $nTiO_2$-A) and alkylalkoxysilane matrices (m-WNC and m-ANC), as reported by the producers: chemical composition, solvent, concentration (w/w) and nanoparticle size (nm).

Component	Description	Solvent	Concentration	NPs Size [1]
$nTiO_2$-W	TiO_2 NPs	water (pH 1.5)	5.5%	50 ± 10
$nTiO_2$-A	TiO_2 NPs	1,2-propanediol	12%	20 ± 5
m-WNC	*n*-propyl trimethoxysilane tris(propyltrimethoxysilyl)amine formic acid	water (pH 4.5)	15%	-
m-ANC	2-methylpropyl trimethoxysilane ethyl orthosilicate butyl orthotitanate (cat.)	2-propanol	40%	-

[1] measured by Dynamic Light Scattering (DLS).

Table 2. Composition (w/w) of WNC and ANC, as reported by the producers.

Product	Solvent	Composition
WNC	water (pH 4.5)	0.96% $nTiO_2$-W 15% m-WNC
ANC	2-propanol	0.12% $nTiO_2$-A 40% m-ANC

Some of the properties of WNC and ANC have been assessed in previous studies [36,37]. The former is a water-based dispersion of alkylalkoxysilane *oligomers* (15% w/w) and TiO_2 nanoparticles (0.96% w/w), while the latter is an alcohol-based solution of alkylalkoxysilane *monomers* (40% w/w) with a lower titania concentration (0.12% w/w). The corresponding TiO_2-free alkylalkoxysilane matrices are hereafter referred to as m-WNC and m-ANC, respectively.

Rheological measurements on the nanocomposites and corresponding matrices were carried out through a Bohlin CV0 120 Rheometer (Bohlin Instruments Vertriebs GmbH, Pforzheim, Germany), using a cone-plate configuration (1° angle, 40 mm diameter) with 0.03 mm gap. Flow curves were recorded for 3 min under 0.1 to 3 Pa stress at 20 °C. Since the fluids exhibited a shear-thinning behaviour, the value of viscosity measured in the low shear rate region around 10 s^{-1} was assumed to be representative of the rheological behaviour of the products in a capillary flow regime. Two measurements were performed for each product. Particle size was measured on a 90 Plus Dynamic Laser Light Scattering instrument (Brookhaven Instruments Corporation, Holtsville, NY, USA) equipped with a 35-mW Laser and an Avalanche photodiode detector collecting the scattered light at 90°. Three measurements were performed for each product.

In order to assess the crystalline form of TiO$_2$ nanoparticles and to investigate their interaction with alkylalkoxysilane matrices upon curing, Raman spectra of TiO$_2$ dispersions (nTiO$_2$-W and nTiO$_2$-A, the latter being previously diluted in 2-propanol), alkylalkoxysilane matrices (m-WNC and m-ANC) and different combinations of them (TiO$_2$: alkylalkoxysilane in 10:1, 2:1, 1:1, 1:5 *w/w* ratio) were recorded after keeping the sols in open vials under controlled humidity and temperature (50% RH and 23 °C) for 40 days until solvent evaporation and curing. Raman spectra of these samples were recorded using a Horiba Jobin Yvon Labram HR800 Raman spectrometer coupled with an Olympus BX41 microscope. The 514.5 nm excitation laser line (Ar$^+$ Sabilite 2017 Spectra-Physics) at 2 mW power was focused by a 50X objective directly on samples deposited on a glass slide. Spectra were recorded by collecting 4 scans and integrating over 10 s. A baseline correction and a smoothing of signal were performed through OMNIC software (Thermo Fischer Scientific, Waltham, MA, USA).

Freshly quarried specimens (5 × 5 × 1 cm^3 and 5 × 5 × 2 cm^3 prisms, 2 and 3 for each size respectively) of Ajarte and Obernkirchen stones were gently polished with abrasive paper (P180 carborundum paper), washed and kept in deionized water for 1 h in order to remove any excess soluble salts. Afterwards, they were dried in oven at 65 °C until constant weight and stored in a silica gel desiccator for 24 h. The products and their respective matrices were applied by capillary absorption for 6 h, using a filter paper pad saturated with the treatments, according to EN standard [38]. After the application, the stone specimens were kept sheltered from direct light for 30 days at the temperature and humidity conditions of the lab (about 23 °C and 50% RH) in order to allow solvent evaporation and the curing of matrices. In order to determine the amount of absorbed product, stone specimens were weighed before and after treatment. The weights were divided by the respective product densities and by the areas of treated surfaces, yielding the volumes of liquid absorbed per unit area (μL/cm^2). Both products and matrices were also cast on 2 glass slides (25 × 75 mm) that had been previously treated with hot Piranha solution (conc. H$_2$SO$_4$ and 30% *w/w* H$_2$O$_2$ in 3:1 volume ratio) for 15 min in order to increase the amount of reactive silanol groups on the glass surface. The slides were kept under saturated solvent (water/2-propanol) atmosphere until complete evaporation of the solvent and formation of a thin film and then stored in a closed vessel for 1 month in order to allow proper curing of the alkylalkoxysilane matrices.

The aesthetic compatibility of treatments was assessed through diffuse reflectance Vis-light spectroscopy (Konica Minolta CM-600D Vis spectrophotometer with a D65 illuminant at 8°, 360–740 nm wavelength range). 25 measurements were performed on each stone specimen before and after the application of treatments according to the EN standard protocol [39]. The results were expressed in the CIE L*a*b* colour space and the average values of L*, a* and b* were used to calculate the colour change ΔE^* according to the formula $\Delta E^* = [(L^*_t - L^*_{nt})^2 + (a^*_t - a^*_{nt})^2 + (b^*_t - b^*_{nt})^2]^{1/2}$, where the subscripts t and nt stand for treated and untreated specimen, respectively. ΔE^* values should not exceed the threshold value of 5 in order for a product to meet the aesthetic requirements for application in the cultural heritage field [40].

The surface morphology of stone specimens was characterized through Environmental Scanning Electron Microscopy (Zeiss EVO 50 EP) before and after the application of treatments. For the Atomic Force Microscopy (AFM) analysis of surface roughness, a Solver Pro AFM microscope (NT-MDT Spectrum Instruments, Beijing, China) was employed, using a silicon cantilever with a tip of 14–16 μm height (NSG10, NT-MDT) and tip curvature radius of 10 nm at a resonant frequency of 140–390 KHz. Measurements were performed in tapping mode at 0.6 Hz scan rate, with 2 scans of a 0.5 × 0.5 μm^2

area. The acquired images were elaborated through the Nova SPM software (NT-MDT Spectrum Instruments). 4 specimens per lithotype were analysed before and after treatment (one specimen for each treatment).

Capillary water absorption was measured on untreated specimens and then after 1 and 2 months from the application of protective treatments according to EN standard [41]. All reported data refer to this latter set of measurements. The specimens were weighed at the following time intervals: 10 min, 20 min, 30 min, 60 min, 4 h, 6 h, 24 h, 48 h, 72 h and 96 h. The area under the absorption curve was calculated through numerical integration. The Relative Capillary Index (CI_{rel}) was used to evaluate the behaviour of treated specimens for the duration of the experiment (96 h), while the Sorptivity (AC) was used to evaluate their short-term behaviour (30 min). For both parameters, values lower than 0.2 indicate a good reduction of water absorption [40]. Static contact angle (θ) measurements were performed on 15 points for each specimen, according to EN standard [42], using an OCA (Optical Contact Angle) 20 PLUS (DataPhysics, San Jose, CA, USA). A drop volume of 5 µL was used and measurements were performed 10 s after drop deposition. Drop profiles were analysed according to Laplace-Young theory. HPLC grade water (Chromasolv ® Plus, Sigma Aldrich, St. Louis, MO, USA) was used as the liquid.

The photocatalytic properties of the products (WNC and ANC) were assessed through the Rhodamine discolouration test, according to the procedure reported in a previous work [36]. For a comparison between products and their respective TiO_2-free matrices, the latters' behaviour was tested as well. The irradiation chamber (Suntest XLS+, URAI S.p.A) was equipped with a Xenon arc lamp (NXE 1700, with a cut-off filter for $\lambda < 295$ nm) producing an irradiance of 765 W/m^2 in the 300–800 nm range. The degradation of Rhodamine B was monitored up to 150 min by means of colour measurements. For the assessment of photocatalytic activity, the a* value from each measurement was considered, which represents the red colour component in the CIE Lab colour space. The extent of discolouration (D*) was then evaluated according to the formula D*(%) = (| a*(t) − a*(rB) | / | a*(rB) − a*(0) |)*100, where a*(0) and a*(rB) are the average values of chromatic coordinate a* before and after the application of the colourant solution and a*(t) is the a* value after t minutes of light exposure. Specimens treated with a commercial water-repellent product, Silres BS 290 (Wacker Chemie GmbH, Munich, Germany), based on a mixture of silanes and siloxanes (8% w/w in white spirit), were used as reference owing to their hydrophobic features making the interaction of the colourant solution with stones comparable with that observed on treated specimens. In order to distinguish the effects of photolytic and thermal degradation of Rhodamine from the actual photocatalytic process, the ratios of D* values for specimens treated with WNC, ANC and their respective matrices (D*$_{PRODUCT}$) and for specimens treated with Silres (D*$_{SILRES}$) at 30, 90 and 150 min are reported as parameters for the evaluation of photocatalytic activity. In the case of TiO_2-free matrices, this ratio is clearly expected to approach unity, while for photocatalytic products greater than 1 values should be obtained.

Specimens treated with WNC, ANC, m-WNC and m-ANC (16 specimens per lithotype, including subsets of $5 \times 5 \times 2$ cm^3 and $5 \times 5 \times 1$ cm^3 specimens, previously characterized according to the testing protocol described above) were subjected to a UV ageing procedure in order to assess the chemical stability of the two organosilica gels and investigating the influence of TiO_2 nanoparticles on possible photo-oxidative effects. Products and matrices applied on glass slides also underwent this ageing procedure. The UV ageing was conducted for 600 h in an irradiation chamber (Suntest XLS+, Atlas GmbH, Ganderkesee, Germany) equipped with a Xenon arc lamp (NXE 1700) simulating daylight (cut-off filter for $\lambda < 295$ nm). The irradiance of the lamp was set to 765 W/m^2 in the 300–800 nm range, with an emission of about 65 W/m^2 from 300–400 nm (as reported by the manufacturer), which is close to the hypothetical upper limit of UV irradiation of natural daylight (70 W/m^2) [27]. The temperature of the specimens, measured through a black body reference, was kept at 65 ± 15 °C. This experimental set-up was comparable to others adopted in literature [27,29]. Water absorption and static contact angle measurements were then performed on aged specimens in order to assess the retention of protective effectiveness. The nanocomposites and corresponding matrices cast on glass slides were analysed through FTIR microscopy before and after ageing in order to characterize their chemical modification. µ-FTIR analysis was conducted in attenuated total reflection (ATR) mode (Ge crystal) on a Thermo Nicolet 6700 spectrophotometer coupled to a Thermo Nicolet Continuum FTIR microscope with MCT detector (128 acquisitions, 650–4000 cm^{-1} spectral window, 4 cm^{-1} resolution). Spectra were processed on OMNIC software (Thermo Fischer Scientific): the baseline was corrected and a reduction of noise was performed via the smoothing function. Secondly, for WNC and ANC, 2 further specimens per lithotype were subjected to a rain ageing procedure with the aim of assessing the mechanic stability of TiO_2 nanoparticles under the action of rain wash-out. The resistance to rain wash-out was assessed by a purposely designed rain chamber, equipped with a peristaltic pump (Behr GmbH, New York, NY, USA) and a set of needles (d = 0.2 mm) providing constant dripping of distilled water with a rate of about 82 mm/h onto specimens placed on a rack and tilted by 45° with respect to the horizontal plane. The experimental set-up was in partial agreement with similar experiments reported in literature [28]. The test was conducted in 4 steps of 24 h, each step consisting of 7 h of wetting followed by 17 h of drying at room temperature. Each specimen was subjected to four rain drops and periodically displaced by 0.5 cm along the horizontal plane in order to obtain a more homogeneous distribution of the rain drops over the tested surface. To the same purpose, the specimens were also rotated by 180° after half testing time. Then, in order to assess the retention of TiO_2 nanoparticles on the stone surfaces, the photocatalytic activity was measured again through the Rhodamine discolouration test.

3. Results and Discussion

3.1. Characterization of Materials

3.1.1. Lithotypes

Ajarte and Obernkirchen are highly porous stones (23.5 and 24.1 vol%, respectively) with different mineralogical and microstructural properties [36,37,43]. (These latter properties are recalled in Table S1 in Supplementary Materials). Ajarte has an almost purely calcitic composition (93.4 mol%) and a low average pore diameter (0.17 µm), whereas Obernkirchen consists mainly of quartz (89.6 mol%) and displays a considerably higher mean pore size (0.76 µm). Although both stones exhibit very close values of open porosity, the suction power and velocity of their capillary networks are remarkably different: Ajarte absorbs water to a far greater extent and at a faster rate than Obernkirchen (Figure 2), which is probably due to differences in pore shape and connectivity.

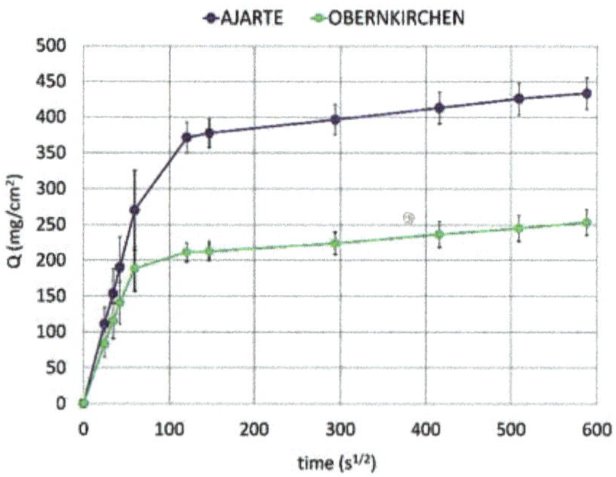

Figure 2. Water absorption by capillarity in untreated lithotypes.

These distinct water absorption regimes will be taken into account when evaluating the penetration of protective treatments into the porous crystalline matrix of the two stones. Furthermore, the different pore-size distributions (see Figure S1 in Supplementary Materials), especially the greater amount of small-size pores found in Ajarte, will be considered as a possible discriminating factor for the behaviour of the products under study, as the latter need to penetrate into the widest possible pore-size range in order to produce a diffuse and effective water-repellent action.

3.1.2. Protective Treatments

The two nanocomposite formulations WNC and ANC display hybrid properties. Upon curing of the alkylalkoxysilane matrices, alkyl groups impart hydrophobic features, while TiO_2 nanoparticles at low concentration are able to provide photocatalytic and self-cleaning properties [36]. The sol-gel condensation of silane precursors upon solvent evaporation and curing produces, to a variable extent, a cross-linked organosilica gel network, in accordance with the usual reactivity of alkylalkoxysilane compounds. Silanol groups in the gels are expected to interact with stone substrates either through condensation with surface silanols of silicate minerals [44] or through the build-up of noncovalent interactions (hydrogen bonding). Furthermore, protonated amine groups (present in WNC) should be involved in ionic interactions with both carbonate and silicate minerals, as it is reported for other aminosilane compounds [45].

Since transport properties have a predictable effect on the penetration of treatments in porous substrates, viscosity and particle size measurements, previously performed on WNC and ANC [37], were also conducted on the corresponding matrices m-WNC and m-ANC in order to assess the modification of these properties upon addition of TiO_2 nanoparticles. The results (Table 3) show that only quite small differences exist in the viscosity of fluids (WNC/m-WNC and ANC/m-ANC), whilst marked differences can be observed among the values of particle size. Actually, it is noteworthy that both m-WNC (which is made up of alkylalkoxysilane oligomers) and the respective TiO_2 nanoparticles (nTiO$_2$-W) consist of aggregates of several tens of nanometres (Tables 1 and 3). Moreover, further aggregation of these silane oligomers and/or TiO_2 nanoparticles probably occurs after the preparation of the mixture, because the final product turns out to have higher particle size compared to its separate components (Table 3). This suggests that the sol increases its aggregation state after the mixing of components and a greater likelihood exists that it will interact with the inner pore surface of stones during the capillary uptake, thus experiencing a more difficult penetration. Conversely, the matrix m-ANC consists of well dispersed alkylalkoxysilane monomers with no light scattering features (Table 3) and its corresponding nanotitania (nTiO$_2$-A) exhibits a low particle size (Table 1). Furthermore, both components do not seem to undergo any aggregation upon mixing, since particles in the formulate (Table 3) are nearly the same size as those of the precursor TiO_2 dispersion. Therefore, given the stability of the sol, which is typical of alcohol-based alkylalkoxysilanes, the product behaves as a non-reactive fluid during the absorption into stones, thus facilitating its own penetration.

Table 3. Main properties of the products (WNC, ANC) and respective matrices (m-WNC and m-ANC): density (g/cm^3), viscosity (mPa·s) and particle size (nm).

	Density	Viscosity	Particle Size
WNC	1.03	10 ± 1	105.9 ± 0.4 [1]
m-WNC	1.03	10 ± 1	82.8 ± 0.2
ANC	0.84	7 ± 1	25 ± 1 [1]
m-ANC	0.84	11 ± 3	-

[1] data taken from [37].

In conclusion, the significantly different particle size and distinct stability of the two nanocomposites towards aggregation are the discriminating factors for their penetration into porous substrates, more so in the case of Ajarte stone whose average pore diameter is around 200 nm (Table S1). In particular, the greater size of WNC particles, which exceeds 100 nm, as well as the lower stability of its components (as it is the case of water-based alkylalkoxysilanes, characterized by a higher reactivity as compared to their solvent-based analogues [46]) can be expected to prevent the penetration of the sol into the thinnest pores and are likely to reduce it for the whole of the other pores. On the other hand, the lower size of TiO_2 nanoparticles in ANC and the stability of its sol components should concur to a greater and more homogeneous penetration of the product into a wider pore size range. From this viewpoint, little difference is to be expected between ANC and its TiO_2-free matrix m-ANC, whereas WNC should penetrate less easily than m-WNC.

Raman analysis of the TiO_2 dispersions ($nTiO_2$-W and $nTiO_2$-A) used to prepare the two formulations was performed after solvent evaporation with the aim of assessing the titania polymorphs present therein. As it is well known, among these polymorphs, pure anatase or anatase in the presence of a small fraction of rutile display the highest photocatalytic activity [47,48]. Raman spectra in Figure 3 reveal that the pattern of $nTiO_2$-W (632, 516 and 407 cm^{-1}), that is, TiO_2 nanoparticles used in WNC formulation, bears a rather good correspondence to that of anatase [49], in spite of peaks at 407 cm^{-1} and 632 cm^{-1} being displaced by about 8 cm^{-1} (upward and downward, respectively) relative to the corresponding signals of the anatase crystal. The presence of some residual rutile phase can also be argued on the basis of the weak signal at 453 cm^{-1} [49]. The observed broadening and peak shift of the Raman signals can be possibly ascribed both to the non-stoichiometry of the samples (i.e., an oxygen deficiency) and to the presence of disorder in TiO_2 nanoaggregates [50]. Moreover, it is well known that confinement effects in nanosized crystalline domains is responsible of a partial relaxation of the Raman selection rules for the crystal, with consequent activation of phonons with wave-vector close to the Γ point in the first Brillouin zone. The above phenomenon is often reported as the main responsible for band broadening, changes of the band shapes and displacements of the band maxima observed in the vibrational spectra of nanocrystals. In the case of $nTiO_2$-A, that is, TiO_2 nanoparticles used in ANC formulation, the interpretation of spectral features turns out to be more complicated. Except for the intense band at 630 cm^{-1}, suggesting the presence of anatase (with the same shift observed for $nTiO_2$-W) and a weaker signal at 524 cm^{-1} that could be given the same interpretation, the third expected signal around 399 cm^{-1} might be identified as the feature observed at 424 cm^{-1}. Furthermore, other signals, not ascribable to either anatase or rutile phase, can be noticed. In particular, the intense band at 838 cm^{-1}, which is due to 1,2-propanediol, indicates that residual solvent is still present in the solid phase, probably adsorbed or covalently bonded [51] to the surface of TiO_2 nanoparticles.

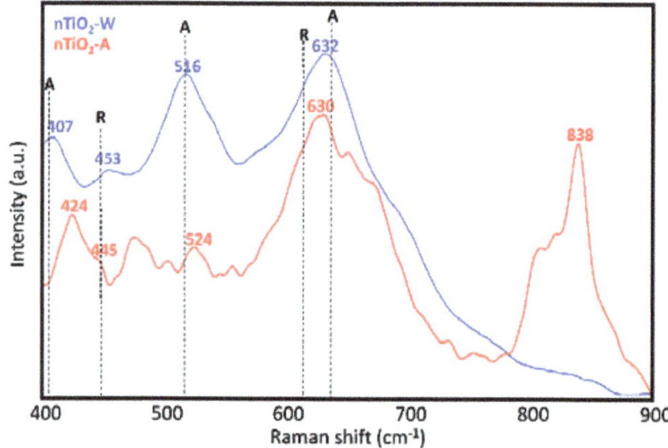

Figure 3. Raman spectra of dry TiO_2 dispersions ($nTiO_2$-W and n-TiO_2-A) with indication of peak positions for anatase (A) and rutile (R) crystals.

In order to assess whether a modification in the chemical environment around TiO_2 nanoparticles occurs by interaction with the organosilica gels upon curing of matrices, several mixtures of TiO_2

and the respective matrices at different weight ratios were also analysed by Raman spectroscopy. In the case of ANC, the Raman study of different mixtures of TiO_2 nanoparticles (nTiO$_2$-A) with the respective matrix (m-ANC) did not allow to detect any new signals ascribable to a chemical interaction between the two components. This agrees with the above made hypothesis that nTiO$_2$-A particles may be actually surrounded by a shell of adsorbed 1,2-propanediol molecules keeping them protected from direct interaction with the organosilica gel. In the case of WNC, changes in the relative intensities and small shifts of several signals of the matrix were observed in the spectra of mixtures, indicating that this matrix undergoes a possibly conformational and/or structural rearrangement upon addition of TiO_2 nanoparticles (Figure 4). Furthermore, it was possible to detect a signal at 1013 cm^{-1}, which can be only observed in the spectra of 1:1 and 1:5 nTiO$_2$-W/m-WNC mixtures, that is, with an excess of silane over TiO_2. According to the literature [52], this signal could be ascribed to the formation of a covalent Si-O-Ti link between TiO_2 and the silanol groups resulting from the hydrolysis of an aminoalkylalkoxysilane compound. The chemical interaction between TiO_2 and the organosilica gel of WNC probably contributes to an increase in the adhesion forces between nanoparticles and the surrounding matrix, thus improving the retention of titania on the treated stone surfaces. This argument will be resumed in the following, while discussing the durability of coatings and, notably, the effects of rain wash-out (Section 3.3).

Then, the comparison between nanocomposites and their respective matrices shows that, while ANC/m-ANC display the same absorption behaviour, the uptake of WNC is considerably lower than that of m-WNC, all of which is once again consistent with the values of particle size reported in Table 3.

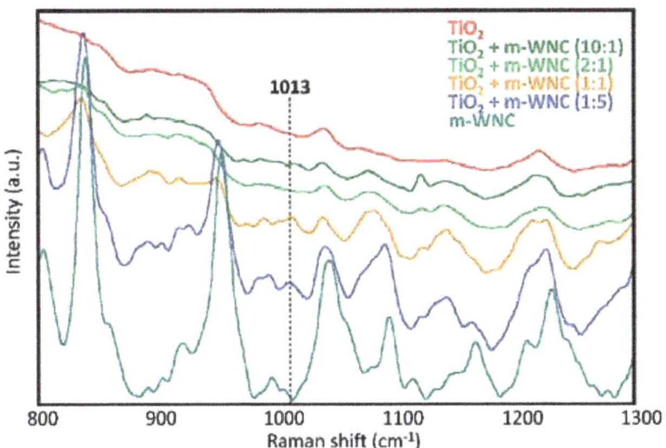

Figure 4. Raman spectra of dry nTiO$_2$-W/m-WNC mixtures at different weight ratios: pure TiO$_2$, 10:1, 2:1, 1:1, 1:5 and pure m-WNC.

3.1.3. Absorption of Protective Treatments

The volumes of WNC, ANC and their respective matrices (m-WNC and m-ANC) absorbed by Ajarte and Obernkirchen (Table 4) indicate that ANC/m-ANC saturate the pore volume of both stones to nearly the same extent as water (Figure 2). This is evidence of the fact that they encounter no difficulty in penetrating into the stone capillary networks, in agreement with their non-aggregating behaviour and low viscosity (Table 3). On the other hand, the absorption of WNC/m-WNC turns out to be much more difficult, most likely due to the higher particle size and greater aggregation of their components. Such effect is especially evident on Ajarte, whose lower average pore diameter and higher pore surface area (Table S1) provide a greater interaction between stone and penetrating fluids and pose higher limitations on the mobility of nanoparticles.

Table 4. Volume of liquid treatments ($\mu L/cm^2$) absorbed by Ajarte and Obernkirchen stones. Values are averaged on 3 specimens.

	WNC	m-WNC	ANC	m-ANC
Ajarte	122 ± 17	203 ± 47	408 ± 6	411 ± 7
Obernkirchen	74 ± 5	213 ± 1	236 ± 19	219 ± 5

3.2. Testing of Treated Lithotypes

3.2.1. Surface Colour Monitoring

Colour measurements (Table 5) indicate that both nanocomposites (WNC and ANC) have good aesthetic compatibility on Ajarte stone, producing colour variations (ΔE^*) well below the threshold value of 5 [41], as it had already been assessed in our previous study [36]. On Obernkirchen, WNC still exhibits a good compatibility, while ANC gives rise to a greater chromatic alteration as a result of darkening ($\Delta L^* = -8$) and yellowing ($\Delta b^* = 5$), which are probably associated with a more extensive coverage of the quartz grains by low surface energy alcohol-based products. The colour analysis of stones treated with m-WNC and m-ANC indicates that nanocomposites behave in much the same way as their respective matrices, hence the addition of TiO_2 nanoparticles does not involve further chromatic variations.

Table 5. Values of ΔE^*, ΔL^*, Δa^* and Δb^* of Ajarte and Obernkirchen stones treated with WNC, ANC and their matrices. Values are averaged on 3 specimens.

		ΔE^*	ΔL^*	Δa^*	Δb^*
AJARTE	WNC	1.5 ± 0.3	-0.9 ± 0.3	-0.36 ± 0.06	1.1 ± 0.3
	m-WNC	1 ± 1	-0.6 ± 0.7	-0.3 ± 0.1	1.1 ± 0.9
	ANC	2.2 ± 0.6	-2.0 ± 0.8	0.21 ± 0.09	-0.3 ± 0.9
	m-ANC	4 ± 2	-3 ± 1	0.3 ± 0.3	1.5 ± 0.6
OBERN.	WNC	2.6 ± 0.6	-2.3 ± 0.5	0.15 ± 0.08	1.1 ± 0.4
	m-WNC	2.5 ± 0.9	-1.7 ± 0.7	0.08 ± 0.09	1.8 ± 0.5
	ANC	10 ± 1	-8 ± 1	1.3 ± 0.3	5 ± 1
	m-ANC	9 ± 1	-7 ± 2	1.0 ± 0.2	4.7 ± 0.6

3.2.2. Evaluation of Surface Morphology

A comparison of the surface morphology of untreated and treated stones through scanning electron microscopy was first carried out (see Figures S2 and S3 in Supplementary Materials). In the case of Ajarte stone, which is characterized by a fine-grained porous microstructure of calcite crystals, the natural stone features are almost completely retained after treatment, even though few localized accumulations of the treatments can be detected. As regards WNC, observations in SE mode allow to detect a diffuse bridging of intercrystalline gaps. In the case of Obernkirchen, that is, a stone characterized by medium-coarse size quartz grains embedded in a very fine-grained silicate cement with evident voids and diffuse network of pores, the coatings can be easily detected even from BSE images. A smoothening of the surface morphology, with covering of the clasts and filling of the surface pores, especially in the case of ANC, can be observed.

AFM measurements of surface roughness (Table 6) were performed to investigate at nanoscale the stone/coating interaction and, in particular, to assess how TiO_2 nanoparticles contribute to the modification of the surface textural properties of stones. The two lithotypes are characterized by different textural features, as Ajarte displays a much greater surface nanoroughness with respect to Obernkirchen. Furthermore, the two alkylalkoxysilane matrices interact with stones in a completely different way: the water-based matrix (m-WNC) reduces surface roughness drastically, as it can be expected from its poorer penetration, more so on Ajarte stone, which exhibits a higher intrinsic nanoroughness. On the other hand, the alcohol-based matrix (m-ANC) tends to enhance the naturally rough topography of stones, notably in the case of Obernkirchen, that is, the stone with lower intrinsic nanoroughness. The addition of titania can be seen to produce different effects depending on the amount of nanoparticles and the properties of the matrix. In ANC, that is, the product with lower titania content (Table 2), the addition of nanoparticles seems to reduce the roughness-inducing effect of the matrix. In the case of WNC, which is characterized by a higher loading of titania and by a matrix with pronounced texture smoothening effects, the presence of nanoparticles can be seen to slightly reduce these effects on Ajarte and results in only a minor increase in surface nanoroughness on Obernkirchen. In conclusion, while the interaction of alkylalkoxysilane matrices with stones is clearly modified by the addition of nanoparticles, there is no clear evidence that the latter contribute to enhancing surface roughness. The lack of this effect is probably due to the comparatively small amount of titania in the two formulations, which is much lower than applied in one previous study where a contribution of nanoparticles to surface roughness was clearly observed [19]. Besides that, it must be considered that on porous stones such contribution is further reduced by the penetration of nanoparticles into the porous matrix.

Table 6. Values of root mean square (RMS) surface Roughness (nm) of Ajarte and Obernkirchen stones: untreated and treated with WNC, m-WNC, ANC and m-ANC.

	Untreated	**WNC**	**m-WNC**	**ANC**	**m-ANC**
Ajarte	33 ± 3	5	1	37	38
Obernkirchen	7 ± 3	12	3	3	39

3.2.3. Evaluation of Water Absorption and Surface Wettability

Our previous studies regarding the protective performance of WNC and ANC [36,37] showed that these products are able to effectively reduce the water uptake both at short-term (AC) and long-term (CI_{rel}) contact, with a slightly higher effectiveness of the alcohol-based product ANC, as can be expected from its higher penetration (Table 4). The results of water absorption measurements by capillarity performed in this study (Table 7) show clearly that nanocomposites (WNC, ANC) and corresponding TiO_2-free matrices (m-WNC, m-ANC) behave in a very similar way, thus proving that the addition of nanoparticles, though reducing the penetration of WNC into the stone pores (Table 4), does not compromise the effectiveness of matrices in protecting stones from water capillary absorption.

Table 7. Amount of water absorbed per unit area at 96 h (Q_i, mg·cm^{-2}) and absorption rate at 30 min (AC, mg·cm^{-2}·s$^{-1/2}$) *before* (nt) and *after* (t) treatment with WNC, m-WNC, ANC and m-ANC for Ajarte and Obernkirchen stones and respective values of Relative Capillary Index (CI_{rel}). Values are averaged on 3 specimens.

		Q_i nt	Q_i t	AC nt	AC t	CI_{rel}
AJARTE	WNC	430 ± 20	79 ± 6	4.2 ± 0.6	0.144 ± 0.007	0.132 ± 0.008
	m-WNC	430 ± 30	79 ± 3	5.2 ± 0.9	0.132 ± 0.001	0.12 ± 0.01
	ANC	447 ± 3	36 ± 3	4.6 ± 0.8	0.112 ± 0.004	0.063 ± 0.005
	m-ANC	438 ± 2	34 ± 4	5.1 ± 0.2	0.106 ± 0.003	0.061 ± 0.008
OBERN.	WNC	257 ± 6	33 ± 5	2.8 ± 0.2	0.076 ± 0.005	0.09 ± 0.01
	m-WNC	250 ± 4	36.4 ± 0.7	3.4 ± 0.4	0.070 ± 0.004	0.099 ± 0.001
	ANC	260 ± 20	18 ± 4	3.7 ± 0.6	0.075 ± 0.006	0.052 ± 0.008
	m-ANC	240 ± 10	14 ± 5	3.2 ± 0.6	0.057 ± 0.005	0.05 ± 0.01

This is a rather interesting result, which suggests that the homogeneity of surface deposition and pore hydrophobization are more critical factors in determining a satisfactory water-barrier effect than is the total amount of applied product. Among the two stones, Ajarte turns out to be harder to protect, which is expected on the basis of its high number of small-size pores (Figure S1) that treatments are less likely to enter and protect.

Both nanocomposites and corresponding matrices are also able to impart high water-repellency (Table 8) to the stone surfaces on which they are applied, yielding contact angles higher than 130°, which amount to so-called "superhydrophobic" behaviour [53]. With the exception of WNC, all treatments give slightly higher contact angles on Ajarte than Obernkirchen, which is consistent with the greater amounts of products absorbed by this stone and its naturally higher surface nanoroughness (Table 6) contributing to water-repellency via the Cassie-Baxter state. As pointed out in the discussion of AFM results (Section 3.2.2), the comparison between nanocomposites and matrices shows that the presence of TiO_2 nanoparticles does not contribute to the water-repellency of matrices through an increase in surface nanoroughness. Actually, it can be observed that ANC gives rise to slightly lower contact angles with respect to its matrix, which can be reasonably ascribed to the higher surface roughness induced by the latter (Table 6). As regards WNC, the nanocomposite behaves in much the same way as m-WNC on Obernkirchen, whereas on Ajarte it induces a contact angle about 8° lower than that provided by the matrix. In this case, the difference is probably due to a less effective surface coverage of one of the three specimens used for the test (as the high standard deviation of measurements indicates), again suggesting that, on a stone characterized by a large number of very small intercrystalline pores, the product is somewhat less effective than its matrix in providing a uniform hydrophobization of the pores' walls due to the larger size of its aggregates (Table 3).

Table 8. Values of static contact angle (θ, °) of water *before* (nt) and *after* (t) treatment with WNC, m-WNC, ANC and m-ANC for Ajarte and Obernkirchen stones. Values for treated stones are averaged on 2 specimens.

		θ nt	θ t
AJARTE	WNC		131 ± 14
	m-WNC	<10 [1]	139 ± 3
	ANC		138 ± 2
	m-ANC		142 ± 4
OBERN.	WNC		140 ± 2
	m-WNC	21 ± 2	138 ± 1
	ANC		133 ± 1
	m-ANC		137 ± 1

[1] contact angles on Ajarte are too low to be measured.

3.2.4. Evaluation of Photocatalytic Activity

In a previous study evaluating the photocatalytic properties of WNC and ANC on Ajarte stone [36], it was shown that specimens treated with WNC exhibit a considerably faster colourant degradation with respect to a non-photocatalytic reference product, thus proving that TiO_2 nanoparticles present in the formulation have a well-defined photocatalytic action. The trend of colourant degradation during the exposure to Xenon lamp irradiation also indicated that the kinetics of photocatalyzed discolouration is very fast within the first 30 min of irradiation and already attains a plateau at 90 min, after which the rate is gradually diminished due to the parallel progress of the slower non-catalysed photo-oxidative reaction. In the case of ANC, a less relevant increase in the discolouration rate was observed, pointing out a lower yet still visible photocatalytic activity. In the present research, analogous results for WNC and ANC are also achieved on Obernkirchen (Table 9), in spite of the different microstructure and mineralogical composition of this stone.

Furthermore, the comparison between nanocomposites and matrices allows to make a more accurate evaluation of the role of TiO_2 nanoparticles in the different rates of discolouration displayed by the two nanocomposites. The higher photocatalytic activity of WNC has been explained by its higher titania content (Table 2) and by the observed tendency of TiO_2 nanoparticles in ANC to aggregate during the curing of the product, thus reducing their specific surface area [36]. The 'hybrid' character of these nanoparticles, which are probably surrounded by a shell of chemisorbed 1,2-propanediol molecules (as discussed in Section 3.1.2), may also contribute to the reduction of their photoactivity.

Table 9. Ratio of discolouration values (D*) of Ajarte and Obernkirchen stones treated with WNC/ANC (D*PRODUCT) and corresponding values for the reference non-photocatalytic product Silres BS 290 (D*SILRES) after 30, 90 and 150 min irradiation.

		D*PRODUCT/D*SILRES		
		30 min	90 min	150 min
AJARTE	WNC [1]	5.6	3.9	3.3
	m-WNC	0.7	0.6	0.6
	ANC [1]	0.2	2.0	2.0
	m-ANC	0.7	1.0	0.8
OBERN.	WNC	4.7	5.2	3.7
	m-WNC	0.8	1.1	0.5
	ANC	2.1	2.2	1.7
	m-ANC	1.1	-	0.7

[1] data taken from [36].

3.3. Evaluation of the Durability of Protective Treatments

The chemical stability of nanocomposites and TiO_2-free matrices upon UV light irradiation was studied by referring to their FTIR spectra, recorded before and after the UV ageing procedure and to the modification of capillary water absorption of stone specimens.

A glance at Figures 5 and 6 shows immediately that WNC/m-WNC have a markedly different behaviour with respect to ANC/m-ANC. While in the first case (Figure 5) spectra before and after ageing show clear differences, which become very impressive in the presence of TiO_2 nanoparticles, the spectra of ANC and m-ANC (Figure 6) after irradiation are practically superimposable to those of the unaged materials. It is worth noticing that in the ANC/m-ANC case, the IR spectrum is quite simple, showing strong bands in the O-H and C-H stretching regions (3500–2800 cm^{-1}), weak features in the 1500–1300 cm^{-1} region and a dominant absorption in the region of Si-O stretching modes (maximum at 1030 cm^{-1}). On the contrary, the spectra of WNC/m-WNC (Figure 5) are characterized by the occurrence of many very strong absorption features, which can be ascribed to the presence of species containing polar groups. In particular, features due to the amine groups are expected based on the material formulation (Table 1). As it will be better analysed in the following discussion, these chemical groups undergo chemical transformations upon irradiation, especially in presence of TiO_2 nanoparticles. In the case of m-WNC, the observed changes of the spectral pattern upon irradiation cannot be ascribed to photo-oxidative degradation, pointing out a rather good chemical stability. In particular, the retention of the C-H stretching band (peaks at 2952–2872 cm^{-1}) and the one due to Si-C stretching at 1230 cm^{-1} indicate that the alkyl moiety of the matrix has not been compromised. The broad feature with maximum at 3330 cm^{-1} can be ascribed to contributions from O-H and N-H stretching modes: Its reduction in intensity can be related to the heat-induced evaporation of water molecules trapped inside the gel matrix, as well as to reactions involving amine groups. For this reason, it is impossible to say whether oxidation products containing hydroxyl groups are formed upon ageing. Other relevant changes in the spectrum are the vanishing of the peak at 1584 cm^{-1} (this feature is compatible with the N-H bending of the amine group, involved in hydrogen bonding [54]), with parallel appearance of one at 1663 cm^{-1}. This last feature could be ascribed to the presence of amide groups, whose formation is consistent with the reactivity displayed by some aminoalkylalkoxysilanes towards carbon dioxide [55]. Actually, by considering that the strong band of the unaged sample at 1584 cm^{-1} shows a shoulder at about 1660 cm^{-1}, it is possible that this peak at 1663 cm^{-1} is already present before irradiation. Quite different from its pure matrix, the ageing of WNC, where TiO_2 is present, leads to the complete mineralization of the organic counterpart and rearrangement of the silica-gel network, suggested by the increase of the O-H stretching band around 3300 cm^{-1}, the appearance of a distinct signal at 1635 cm^{-1} due to O-H bending, the disappearance of the C-H stretching peaks (2872–2952 cm^{-1}) and the change in the shape of the Si-O-Si stretching band around 1110 cm^{-1}. A decrease in the intensity of the TiO$_2$-related band below 700 cm^{-1} can also be observed, yet the form of the band is retained, indicating that TiO$_2$ is still present in the composite. Therefore, the addition of TiO$_2$ nanoparticles at nearly 1% w/w leads to a drastic modification of the ageing resistance of the matrix, accelerating photo-oxidative reactions with consequent predictable loss of its water-repellent properties. The IR analysis of ANC and its matrix upon ageing (Figure 6) shows instead in both cases a complete retention of the original spectral features (with the only exception of a small peak appearing at 932 cm^{-1}, probably due to some modification occurring in the siloxane backbone). In particular, no decrease of C-H related bands and no increase of those related to –OH groups can be observed, hence a photo-oxidative degradation of the alkyl chains responsible for coating's hydrophobicity, either in the pure matrix or in the nanocomposite, can be ruled out. Therefore, in this case, the addition of a lower amount of TiO$_2$, possibly in a less photo-active form (see Section 3.1.2), does not seem to compromise the chemical stability of the matrix.

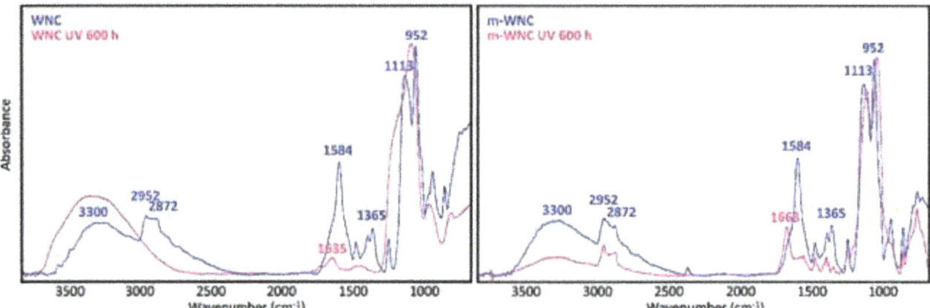

Figure 5. µ-FTIR spectra (ATR mode) of WNC and m-WNC films on glass slides *before* and *after* UV ageing for 600 h.

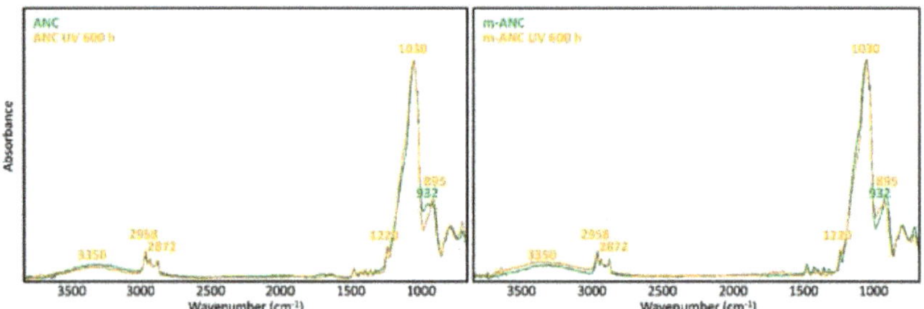

Figure 6. µ-FTIR spectra (ATR mode) of ANC and m-ANC films on glass slides *before* and *after* UV ageing for 600 h.

The trends of change in capillary water absorption for stone specimens treated with WNC, ANC and their respective matrices after UV ageing (Figure 7) are in agreement with the findings of IR analysis. Indeed, while TiO$_2$-free matrices look unaffected by UV light irradiation and retain their good protective performance, the addition of TiO$_2$ nanoparticles in the products leads to different effects, from no increase in water absorption in the case of ANC to a moderate increase in the case of WNC, that is, the product with higher titania content.

On one hand, this confirms that the presence of a higher loading of TiO$_2$ nanoparticles, contributing to the UV-induced degradation of the hydrophobic moiety of the matrix, ends up reducing the "protective ability" of the coating towards the capillary absorption of liquid water. However, from the viewpoint of coating performance, this reduction of protective ability is quite limited, while the fact that the water-barrier effect is not severely compromised proves that the degradative effects do not extend beyond a small depth from the stone surface, thus leaving the bulk of the coating inside the pores largely unaffected. This is a rather significant result, which adds an important piece of information to the knowledge of the durability of photocatalytic coatings applied on stones.

A second, not less important result, comes from the assessment of the mechanic stability of TiO_2 nanoparticles towards rain wash-out, which was only performed on specimens treated with WNC, that is, the product with distinct photocatalytic features (Section 3.2.4). The results of the Rhodamine test performed after ageing (Figure 8) show that a good level of photocatalytic activity is retained on both Ajarte and Obernkirchen stones, indicating the persistence of TiO_2 nanoparticles on the stone surfaces. This is also attested by EDX analysis showing the presence and homogeneous distribution of silicon and titanium. Poor adhesion to stone surfaces is notoriously one of the drawbacks of using bare TiO_2 dispersions and one of the reasons for switching over to nanocomposite formulations. The assessed "physical-mechanical" stability of TiO_2 nanoparticles in WNC is an argument in support of the claim that a good adhesion of TiO_2 to stone is achieved through the embedding organosilica gel matrix. The organosilica gel of WNC, which is based on an aminoalkylalkoxysilane precursor (Table 1), is expected to have good adhesion properties towards both calcareous and siliceous stones. Besides that, as the interpretation of Raman spectra suggested (Section 3.1.2.), a good interaction is achieved between TiO_2 nanoparticles and the matrix, probably resulting in a greater stability of the former inside the protective coating.

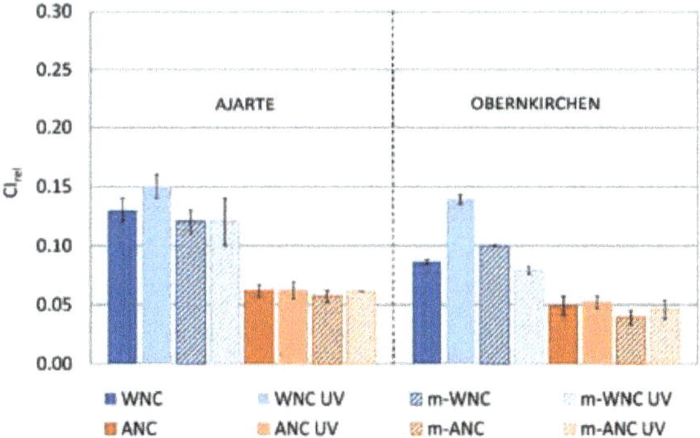

Figure 7. CI_{rel} values of Ajarte and Obernkirchen stones treated with WNC, ANC, m-WNC and m-ANC *before* and *after* UV ageing for 600 h.

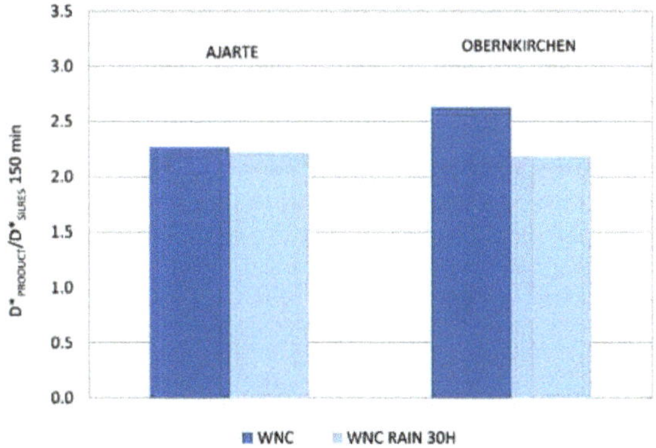

Figure 8. Ratio of discolouration values (D*) of WNC ($D*_{PRODUCT}$) and corresponding values for the reference non-photocatalytic product Silres BS 290 ($D*_{SILRES}$) at 150 min irradiation for Ajarte and Obernkirchen stones after rain ageing for 30 h.

4. Conclusions

This study considered the role of TiO_2 nanoparticles in the stone/coating interaction and the durability of two photocatalytic nanocomposites for the protection of stones, which rely on the combination of alkylalkoxysilane matrices (water- or alcohol-based) with different TiO_2 nanoparticles. The specific aims of the research were to assess how TiO_2 nanoparticles interact with matrices and how they modify the latters' interaction with stones and the durability of the coatings. To address these aims, the two nanocomposite formulations were compared to the respective TiO_2-free matrices and their protective behaviour was assessed on two porous stones with different microstructural and mineralogical properties. The following results were achieved:

- The aggregate size and reactivity of the nanocomposite formulations and the mean pore diameter of stones turned out to be the most relevant factors determining their different absorption and penetration. The addition of TiO_2 nanoparticles to alkylalkoxysilane matrices was shown to produce different effects depending on the reactivity of the matrix and the amount of nanoparticles. For the less reactive alcohol-based matrix (m-ANC), the nanocomposite retains the penetration ability and the protective properties of the silane precursor. For the more reactive water-based matrix (m-WNC) an effect on the aggregation state of the alkylalkoxysilane component was observed, resulting in a lower penetration of the composite product.

- Despite its lower penetration, the water-based nanocomposite WNC showed a good protective performance, particularly on the stone with higher mean pore diameter (Obernkirchen), indicating that the homogeneity of surface deposition and pore hydrophobization are more critical factors in determining a good water-repellency than is the total amount of applied product. The addition of nanoTiO$_2$ to m-WNC did not modify the protective effectiveness of the matrix, whilst it gave rise to the desired significant photocatalytic activity. In the case of the alcohol-based nanocomposite ANC, in spite of the good penetration of the treatment and very good protective performance, the reduced amount of TiO_2 nanoparticles did not allow to obtain distinct photocatalytic features. Moreover, this product caused a visible chromatic alteration on the siliceous stone (Obernkirchen) and in real conditions it is highly preferred to work with water-based formulations.

- The investigation of the durability of coatings upon exposure to UV light clarified that TiO_2 nanoparticles, at the higher concentration found in WNC, contribute to a photo-induced oxidative degradation of the organic component of the matrix, revealed by FTIR analysis. Nevertheless, an important result of this study was that this degradation does not substantially compromise the effectiveness of the coating in reducing water capillary absorption, because it does not extend to the pore network under the surface where the coating imparts most of its water-barrier effect. Furthermore, the good interaction achieved between TiO_2 nanoparticles and the embedding matrix in WNC, assessed through Raman analysis, resulted in a stable anchoring of nanoparticles to the stone surfaces even after prolonged exposure to in-lab simulated rain wash-out.

As a complementary part to this study, the nanocomposite coatings have also been applied in situ on small pilot areas of the Cathedrals of Vitoria-Gasteiz and Cologne, where the selected stones are used as building materials and a long-term on-site monitoring of their performance is currently underway in order to examine their durability under real environmental conditions.

Supplementary Materials: The following are available online at http://www.mdpi.com/1996-1944/11/11/2289/s1. Table S1: Microstructural features of lithotypes: total open porosity (vol%), average pore diameter (µm), pore surface area (m^2/g) and bulk density (g/cm^3); Figure S1: Pore-size distribution of lithotypes; Figure S2: SEM images of Ajarte stone in BSE (left) and SE (right) mode: Untreated (A) and treated with WNC (B), m-WNC (C), ANC (D), m-ANC (E); Figure S3: SEM images of Obernkirchen stone in BSE (left) and SE (right) mode: Untreated (A) and treated with WNC (B), m-WNC (C), ANC (D), m-ANC (E).

Author Contributions: Conceptualization, F.G. and L.T.; methodology, F.G. and M.R.; investigation, M.R, F.G. and L.B.; resources, L.T.; data elaboration, M.R., F.G. and L.B.; writing—original draft preparation, M.R.; writing—review and editing, F.G., C.C and L.T.; project administration, L.T.; funding acquisition, L.T.

Funding: This research was funded by the EU Horizon 2020 programme, grant number 646178 ("Nano-Cathedral-Nanomaterials for conservation of European architectural heritage developed by research on characteristic lithotypes").

Acknowledgments: The authors gratefully acknowledge the collaboration of ChemSpec srl (Italy) and Colorobbia Consulting srl (Italy) for supplying the nanomaterials and the related information.

Conflicts of Interest: The authors declare no conflict of interest.

References

1. Lackhoff, M.; Prieto, X.; Nestle, N.; Dehn, F.; Niessner, R. Photocatalytic activity of semiconductor-modified cement—Influence of semiconductor type and cement ageing. *Appl. Catal. B* **2003**, *43*, 205–216. [CrossRef]

2. Chen, J.; Poon, C.-S. Photocatalytic construction and building materials: From fundamentals to applications. *Build. Environ.* **2009**, *44*, 1899–1906. [CrossRef]

3. Quagliarini, E.; Bondioli, F.; Goffredo, G.B.; Licciulli, A.; Munafò, P. Smart surfaces for architectural heritage: Preliminary results about the application of TiO_2-based coatings on travertine. *J. Cult. Herit.* **2012**, *13*, 204–209. [CrossRef]

4. Bergamonti, L.; Alfieri, I.; Lorenzi, A.; Montenero, A.; Predieri, G.; Barone, G.; Mazzoleni, P.; Pasquale, S.; Lottici, P.P. Nanocrystalline TiO_2 by sol–gel: Characterisation and photocatalytic activity on Modica and Comiso stones. *Appl. Surf. Sci.* **2013**, *282*, 165–173. [CrossRef]

5. Munafò, P.; Goffredo, G.B.; Quagliarini, E. TiO_2-based nanocoatings for preserving architectural stone surfaces: An overview. *Constr. Build. Mater.* **2015**, *84*, 201–218. [CrossRef]

6. Quagliarini, E.; Graziani, L.; Diso, D.; Licciulli, A.; D'Orazio, M. Is nano-TiO_2 alone an effective strategy for the maintenance of stones in cultural heritage? *J. Cult. Herit.* **2018**, *30*, 81–91. [CrossRef]

7. Toniolo, L.; Gherardi, F. The protection of marble surfaces: The challenge to develop suitable nanostructured treatments. In *Advanced Materials for the Conservation of Stone*, 1st ed.; Hosseini, M., Karapanagiotis, I., Eds.; Springer: New York, NY, USA, 2017; Volume 1, pp. 57–78. ISBN 978-3-319-72259-7.

8. Manoudis, P.N.; Tsakalof, A.; Karapanagiotis, I.; Zuburtikudis, I.; Panayiotou, C. Fabrication of super-hydrophobic surfaces for enhanced stone protection. *Surf. Coat. Technol.* **2009**, *203*, 1322–1328. [CrossRef]

9. De Ferri, L.; Lottici, P.P.; Lorenzi, A.; Montenero, A.; Salvioli-Mariani, E. Study of silica nanoparticles—Polysiloxane hydrophobic treatments for stone-based monument protection. *J. Cult. Herit.* **2011**, *12*, 356–363. [CrossRef]

10. Pinho, L.; Mosquera, M.J. Titania-silica nanocomposite photocatalysts with application in stone self-cleaning. *J. Phys. Chem. C* **2011**, *115*, 22851–22862. [CrossRef]

11. Pinho, L.; Elhaddad, F.; Facio, D.S.; Mosquera, M.J. A novel TiO_2–SiO_2 nanocomposite converts a very friable stone into a self-cleaning building material. *Appl. Surf. Sci.* **2013**, *275*, 389–396. [CrossRef]

12. Kapridaki, C.; Pinho, L.; Mosquera, M.J.; Maravelaki-Kalaitzaki, P. Producing photoactive, transparent and hydrophobic SiO_2-crystalline TiO_2 nanocomposites at ambient conditions with application as self-cleaning coatings. *Appl. Catal. B* **2014**, *156–157*, 416–427. [CrossRef]

13. Kapridaki, C.; Maravelaki-Kalaitzaki, P. TiO_2-SiO_2-PDMS nano-composite hydrophobic coating with self-cleaning properties for marble protection. *Prog. Org. Coat.* **2013**, *76*, 400–410. [CrossRef]

14. Pinho, L.; Rojas, M.; Mosquera, M.J. Ag-SiO_2-TiO_2 nanocomposite coatings with enhanced photoactivity for self-cleaning application on building materials. *Appl. Catal. B* **2015**, *178*, 144–154. [CrossRef]

15. Gherardi, F.; Goidanich, S.; Dal Santo, V.; Toniolo, L. Layered Nano-TiO_2 Based Treatments for the Maintenance of Natural Stones in Historical Architecture. *Angew. Chem. Int. Ed.* **2018**, *57*, 7360–7363. [CrossRef] [PubMed]

16. Ksinopoulou, E.; Bakolas, A.; Moropoulou, A. Modifying Si-based consolidants through the addition of colloidal nano-particles. *Appl. Phys. A* **2016**, *122*, 267. [CrossRef]

17. Aslanidou, D.; Karapanagiotis, I.; Lampakis, D. Waterborne superhydrophobic and superoleophobic coatings for the protection of marble and sandstone. *Materials* **2018**, *11*, 585. [CrossRef] [PubMed]

18. Frigione, M.E.; Lettieri, M.T. Novel Attribute of Organic–Inorganic Hybrid Coatings for Protection and Preservation of Materials (Stone and Wood) Belonging to Cultural Heritage. *Coatings* **2018**, *8*, 319. [CrossRef]

19. Gherardi, F.; Goidanich, S.; Toniolo, L. Improvements in marble protection by means of innovative photocatalytic nanocomposites. *Prog. Org. Coat.* **2018**, *121*, 13–22. [CrossRef]

20. Colangiuli, D.; Calia, A.; Bianco, N. Novel multifunctional coatings with photocatalytic and hydrophobic properties for the preservation of the stone building heritage. *Constr. Build. Mater.* **2015**, *93*, 189–196. [CrossRef]

21. Cappelletti, G.; Fermo, P.; Camiloni, M. Smart hybrid coatings for natural stones conservation. *Prog. Org. Coat.* **2015**, *78*, 511–516. [CrossRef]

22. La Russa, M.F.; Rovella, N.; Alvarez De Buergo, M.; Belfiore, C.M.; Pezzino, A.; Crisci, G.M.; Ruffolo, S.A. Nano-TiO$_2$ coatings for cultural heritage protection: The role of the binder on hydrophobic and self-cleaning efficacy. *Prog. Org. Coat.* **2016**, *91*, 1–8. [CrossRef]

23. Milanesi, F.; Cappelletti, G.; Annunziata, R.; Bianchi, C.L.; Meroni, D.; Ardizzone, S. Siloxane-TiO$_2$ hybrid nanocomposites. The structure of the hydrophobic layer. *J. Phys. Chem. C* **2010**, *114*, 8287–8293. [CrossRef]

24. Spinella, A.; Bondioli, F.; Nasillo, G.; Renda, V.; Caponetti, E.; Messori, M.; Morselli, D. Organic-inorganic nanocomposites prepared by reactive suspension method: Investigation on filler/matrix interactions and their effect on the nanoparticles dispersion. *Colloid Polym. Sci.* **2017**, *295*, 695–701. [CrossRef]

25. Gherardi, F.; Gulotta, D.; Goidanich, S.; Colombo, A.; Toniolo, L. On-site monitoring of the performance of innovative treatments for marble conservation in architectural heritage. *Herit. Sci.* **2017**, *5*, 1–15. [CrossRef]

26. Carmona-Quiroga, P.M.; Martínez-Ramírez, S.; Viles, H.A. Efficiency and durability of a self-cleaning coating on concrete and stones under both natural and artificial ageing trials. *Appl. Surf. Sci.* **2018**, *433*, 312–320. [CrossRef]

27. Munafò, P.; Quagliarini, E.; Goffredo, G.B.; Bondioli, F.; Licciulli, A. Durability of nano-engineered TiO$_2$ self-cleaning treatments on limestone. *Constr. Build. Mater.* **2014**, *65*, 218–231. [CrossRef]

28. Franzoni, E.; Fregni, A.; Gabrielli, R.; Graziani, G.; Sassoni, E. Compatibility of photocatalytic TiO$_2$-based finishing for renders in architectural restoration: A preliminary study. *Build. Environ.* **2014**, *80*, 125–135. [CrossRef]

29. Graziani, L.; Quagliarini, E.; Bondioli, F.; D'Orazio, M. Durability of self-cleaning TiO$_2$ coatings on fired clay brick façades: Effects of UV exposure and wet & dry cycles. *Build. Environ.* **2014**, *71*, 193–203. [CrossRef]

30. Maury-Ramirez, A.; Demeestere, K.; De Belie, N. Photocatalytic activity of titanium dioxide nanoparticle coatings applied on autoclaved aerated concrete: Effect of weathering on coating physical characteristics and gaseous toluene removal. *J. Hazard. Mater.* **2012**, *211–212*, 218–225. [CrossRef] [PubMed]

31. Aldoasri, M.A.; Darwish, S.S.; Adam, M.A.; Elmarzugi, N.A.; Ahmed, S.M. Protecting of marble stone facades of historic buildings using multifunctional TiO$_2$ nanocoatings. *Sustainability* **2017**, *9*, 2002. [CrossRef]

32. Goffredo, G.B.; Terlizzi, V.; Munafò, P. Multifunctional TiO$_2$-based hybrid coatings on limestone: Initial performances and durability over time. *J. Build. Eng.* **2017**, *14*, 134–149. [CrossRef]

33. Calia, A.; Lettieri, M.; Masieri, M. Durability assessment of nanostructured TiO$_2$ coatings applied on limestones to enhance building surface with self-cleaning ability. *Build. Environ.* **2016**, *110*, 1–10. [CrossRef]

34. Scalarone, D.; Lazzari, M.; Chiantore, O. Acrylic protective coatings modified with titanium dioxide nanoparticles: Comparative study of stability under irradiation. *Polym. Degrad. Stab.* **2012**, *97*, 2136–2142. [CrossRef]

35. Lazzeri, A.; Coltelli, M.B.; Castelvetro, V.; Bianchi, S.; Chiantore, O.; Lezzerini, M.; Niccolai, L.; Weber, J.; Rohatsch, A.; Gherardi, F.; et al. European Project NANO-CATHEDRAL: Nanomaterials for conservation of European architectural heritage developed by research on characteristic lithotypes. In *Science and Art: A Future for Stone, Proceedings of the 13th International Congress on the Deterioration and Conservation of Stone, Glasgow, UK, 6–10 September 2016*; Hughes, J., Howind, T., Eds.; University of the West of Scotland: Paisley, UK, 2016; Volume 1.

36. Gherardi, F.; Roveri, M.; Goidanich, S.; Toniolo, L. Photocatalytic Nanocomposites for the Protection of European Architectural Heritage. *Materials* **2018**, *11*, 65. [CrossRef] [PubMed]

37. Roveri, M.; Raneri, S.; Bianchi, S.; Gherardi, F.; Castelvetro, V.; Toniolo, L. Electrokinetic Characterization of Natural Stones Coated with Nanocomposites for the Protection of Cultural Heritage. *Appl. Sci.* **2018**, *8*, 1694. [CrossRef]

38. European Committee for Standardization. *EN 16581:2014. Conservation of Cultural Heritage—Surface Protection for Porous Inorganic Materials—Laboratory Test Methods for the Evaluation of the Performance of Water Repellent Products*; European Committee for Standardization: Brussels, Belgium, 2014.

39. European Committee for Standardization. *EN 15886:2000. Conservation of Cultural Property—Test Methods—Colour Measurements of Surfaces*; European Committee for Standardization: Brussels, Belgium, 2000.

40. García, O.; Malaga, K. Definition of the procedure to determine the suitability and durability of an anti-graffiti product for application on cultural heritage porous materials. *J. Cult. Herit.* **2012**, *13*, 77–82. [CrossRef]

41. European Committee for Standardization. *EN 15801:2009. Conservation of Cultural Property—Test Methods—Determination of Water Absorption by Capillarity*; European Committee for Standardization: Brussels, Belgium, 2009.

42. European Committee for Standardization. *EN 15802:2009. Conservation of Cultural Property—Test Methods—Determination of Static Contact Angle*; European Committee for Standardization: Brussels, Belgium, 2009.

43. Lezzerini, M.; Marroni, M.; Raneri, S.; Tamayo, S.; Narbona, B.; Fernández, B.; Weber, J.; Ghaffari, E.; Ban, M.; Rohatsch, A. *D1.5—Mapping of Stones and Their Decay: Part I—Natural Stone Test Methods*; NanoCathedral Project Grant Agreement No. 646178—Confidential Deliverable; Nano-Cathedral: Pisa, Italy, 2017; pp. 8–16.

44. Demjén, Z.; Purkánszky, B.; Földes, E.; Nagy, J. Interaction of silane coupling agents with $CaCO_3$. *J. Colloid Interface Sci.* **1997**, *190*, 427–436. [CrossRef]

45. Kim, J.; Seidler, P.; Wan, L.S.; Fill, C. Formation, structure, and reactivity of amino-terminated organic films on silicon substrates. *J. Colloid Interface Sci.* **2009**, *329*, 114–119. [CrossRef] [PubMed]

46. Charola, A.E. Water repellents and other "protective" treatments: A critical review. In Proceedings of the Hydrophobe III—3rd International Conference on Surface Technology with Water Repellent Agents, Hannover, Germany, 25–26 September 2001; Aedificatio Publishers: Hannover, Germany, 2001.

47. Tanaka, K.; Capule, M.F.V.; Hisanaga, T. Effect of crystallinity of TiO_2 on its photocatalytic action. *Chem. Phys. Lett.* **1991**, *187*, 73–76. [CrossRef]

48. Ding, Z.; Lu, G.Q.; Greenfield, P.F. Role of the Crystallite Phase of TiO_2 in Heterogeneous Photocatalysis for Phenol Oxidation in Water. *J. Phys. Chem. B* **2000**, *104*, 4815–4820. [CrossRef]

49. Burgio, L.; Clark, R.J.H. Library of FT-Raman spectra of pigments, minerals, pigment media and varnishes, and supplement to existing library of Raman spectra of pigments with visible excitation. *Spectrochim. Acta Part A* **2001**, *57*, 1491–1521. [CrossRef]

50. Venkatasubbu, G.D.; Ramakrishnan, V.; Sasirekha, V.; Ramasamy, S.; Kumar, J. Influence of particle size on the phonon confinement of TiO_2 nanoparticles. *J. Exp. Nanosci.* **2014**, *9*, 661–668. [CrossRef]

51. Morselli, D.; Niederberger, M.; Bilecka, I.; Bondioli, F. Double role of polyethylene glycol in the microwaves-assisted non-hydrolytic synthesis of nanometric TiO_2: Oxygen source and stabilizing agent. *J. Nanopart. Res.* **2014**, *16*, 4519–4521. [CrossRef]

52. Meroni, D.; Lo Presti, L.; Di Liberto, G.; Ceotto, M.; Acres, R.G.; Prince, K.C.; Bellani, R.; Soliveri, G.; Ardizzone, S. A close look at the structure of the TiO_2-APTES interface in hybrid nanomaterials and its degradation pathway: An experimental and theoretical study. *J. Phys. Chem. C* **2017**, *121*, 430–440. [CrossRef] [PubMed]

53. Manoudis, P.N.; Karapanagiotis, I.; Tsakalof, A.; Zuburtikudis, I.; Panayiotou, C. Superhydrophobic composite films produced on various substrates. *Langmuir* **2008**, *24*, 11225–11232. [CrossRef] [PubMed]

54. Peña-Alonso, R.; Rubio, F.; Rubio, J.; Oteo, J.L. Study of the hydrolysis and condensation of γ-Aminopropyltriethoxysilane by FT-IR spectroscopy. *J. Mater. Sci.* **2007**, *42*, 595–603. [CrossRef]

55. Blasucci, V.; Dilek, C.; Huttenhower, H.; John, E.; Llopis-Mestre, V.; Pollet, P.; Eckert, C.A.; Liotta, C.L. One-component, switchable ionic liquids derived from siloxylated amines. *Chem. Commun.* **2009**, *1*, 116–118. [CrossRef] [PubMed]

© 2018 by the authors. Licensee MDPI, Basel, Switzerland. This article is an open access article distributed under the terms and conditions of the Creative Commons Attribution (CC BY) license (http://creativecommons.org/licenses/by/4.0/).

 *materials*

Article

Facile Synthesis and Characterization of Ag$_3$PO$_4$ Microparticles for Degradation of Organic Dyestuffs under White-Light Light-Emitting-Diode Irradiation

Chi-Shun Tseng [1], Tsunghsueh Wu [2,*] and Yang-Wei Lin [1,*]

[1] Department of Chemistry, National Changhua University of Education, Changhua City 500, Taiwan; sars124578@yahoo.com.tw

[2] Department of Chemistry, University of Wisconsin-Platteville, 1 University Plaza, Platteville, WI 53818-3099, USA

* Correspondence: wut@uwplatt.edu (T.W.); linywjerry@cc.ncue.edu.tw (Y.-W. L.); Tel.: +1-608-342-6018 (T.W.); +886-4-723-2105-3553 (Y.-W. L.)

Received: 13 April 2018; Accepted: 28 April 2018; Published: 30 April 2018

Abstract: This study demonstrated facile synthesis of silver phosphate (Ag$_3$PO$_4$) photocatalysts for the degradation of organic contaminants. Ag$_3$PO$_4$ microparticles from different concentrations of precursor, AgNO$_3$, were produced and characterized by scanning electron microscopy, powder X-ray diffraction, and UV–visible diffuse reflectance spectroscopy. Degradation rates of methylene blue (MB) and phenol were measured in the presence of microparticles under low-power white-light light-emitting-diode (LED) irradiation and the reaction rate followed pseudo-first-order kinetics. The prepared Ag$_3$PO$_4$ microparticles displayed considerably high photocatalytic activity (>99.8% degradation within 10 min). This can be attributed to the microparticles' large surface area, the low recombination rate of electron–hole pairs and the higher charge separation efficiency. The practicality of the Ag$_3$PO$_4$ microparticles was validated by the degradation of MB, methyl red, acid blue 1 and rhodamine B under sunlight in environmental water samples, demonstrating the benefit of the high photocatalytic activity from Ag$_3$PO$_4$ microparticles.

Keywords: hydrothermal synthesis; silver phosphate; degradation; low power white-light LED irradiation

---

1. Introduction

Advanced oxidation processes (AOPs) have gained considerable attention for the treatment of organic dyestuffs polluted environmental water matrices [1–4]. Among the various AOPs, heterogeneous photocatalysis, which employs semiconductor-based photocatalysts, has been the subject of numerous studies on due to its ability to decompose organic compounds under solar light irradiation [5–9]. For example, the heterogeneous photocatalyst TiO$_2$ is considered an excellent material for water treatment because it is cheap, nontoxic and abundant in nature [5,6,10,11]. However, its inefficient photocatalytic degradation and antibacterial activity render it non-ideal for practical applications because of the rapid recombination of its photogenerated electron–hole pairs leading to high operating costs [12]. Furthermore, as a material with wide band gap, the absorption peak of TiO$_2$ is in the ultraviolet (UV) region (3.2 eV), inferring that TiO$_2$ utilizes only a small proportion of solar energy. To overcome the limitation of TiO$_2$, two synthetic strategies have been proposed. One is to modify the narrow the wide band gap of TiO$_2$ by doping it with other materials, but this method often requires complicated synthetic procedure [13–15]. The other is to synthesize novel semiconductor-based photocatalysts, other than TiO$_2$, capable of absorbing visible light with relative simple synthetic route.

In 2010, Yi et al. synthesized Ag_3PO_4 with high photo-oxidative activity for O_2 evolution from water and the decomposition of organic dyes under visible-light irradiation [16]. The photodegradation rate of organic dyes using this novel photocatalyst is 10 folds higher than that using other visible-light-responsive photocatalysts such as $BiVO_4$ and N-doped TiO_2. Since 2010, studies have focused on the synthesis of different-morphology Ag_3PO_4 photocatalysts and the evaluation of their photocatalytic activity [17–26]. For example, Hsieh et al. proposed shape-controllable synthesis of Ag_3PO_4 crystals and examined their facet-dependent optical properties, photocatalytic activity and electrical conductivity [27]. Ag_3PO_4 cubes have shown more photocatalytically active than rhombic dodecahedra at degrading methyl orange whereas tetrahedral Ag_3PO_4 is inactive. Furthermore, the effects that synthesis temperature has on the morphology and photocatalytic performance of Ag_3PO_4 were demonstrated by Song [28]. They concluded that increasing the synthesis temperature to 120 °C increased the ratio of exposed (110) facets and produced crystals with higher photocatalytic efficiency over rhodamine B under visible-light irradiation. Another study focused on the effects of annealing temperature and time on the structure, morphology and photocatalytic activity of Ag_3PO_4 microparticles [29]. The results revealed that neither the crystalline phase nor photocatalytic activity decreased with annealing temperature, even when the Ag_3PO_4 powders were annealed at 400 °C for 30 min. The excellent photocatalytic activity of the sintered samples in that study were speculated to arise from high crystal quality. Although Ag_3PO_4 is a promising candidate for use in environmental remediation, the large amount of Ag required to produce it and its low structural stability strongly limit its practical applications [30–32].

In the present study, we examined the photocatalytic activity of Ag_3PO_4 with different amounts of $AgNO_3$ through a sonochemical (20 kHz, 600 W for 30 min) and hydrothermal method (180 °C for 24 h). The photocatalytic degradation activity of the as-prepared Ag_3PO_4 microparticles was evaluated according to their degradation of methylene blue (MB) and phenol under low-power white-light light-emitting-diode (LED) irradiation (5 W). Finally, we propose the possible photodegradation mechanism by using the Ag_3PO_4 microparticles.

2. Materials and Methods

2.1. Chemicals

All chemicals were of analytical grade and the highest purity available. $AgNO_3$, Na_3PO_4, HNO_3, phenol, p-benzoquinone, tert-butanol (t-BuOH), and ethylenediamine tetraacetate (EDTA) were purchased from Sigma-Aldrich (St. Louis, MO, USA). MB, methyl red (MR), acid blue 1, and rhodamine B (RhB) were purchased from Invitrogen (Eugene, OR, USA). Milli-Q deionized water was used to prepare all solutions in all experiments.

2.2. Preparation

Ag_3PO_4 microparticles (S1–S4) were prepared through a sonochemical and hydrothermal method. In a typical synthesis procedure, $AgNO_3$ (S1: 1.25, S2: 7.5, S3: 12.5, and S4: 25.0 mmol) was dissolved in 45 mL of deionized water. The prepared Na_3PO_4 solution (2.50 mmol, 5 mL) was then added to the preceding solution under ultrasonic irradiation at room temperature. After 30 min of ultrasonic irradiation (Misonix, Inc., Farmingdale, NY, USA: XL-2020, 20 kHz, 600 W), the mixture was transferred into a Teflon-lined stainless steel autoclave. The autoclave was sealed and heated in an electric oven at 180 °C for 24 h. After the autoclave naturally cooled to room temperature, the precipitates were centrifuged and washed three times with ethanol and deionized water, and then dried in vacuum at 50 °C for 8 h.

2.3. Characterization

The crystalline phases of the Ag_3PO_4 microparticles were determined through X-ray diffraction (XRD) using a SMART APEX II X-ray diffractometer (Bruker AXS, Billerica, MA, USA) with

Cu Kα radiation (λ = 0.15418 nm). The morphologies of the Ag_3PO_4 microparticles were observed through a HITACHI S-4800 scanning electron microscope (SEM) (Hitachi high technologies corporation, Tokyo, Japan) operating at 15 kV. UV–visible diffuse reflectance spectra (UV-Vis-DRS) of the Ag_3PO_4 microparticles were collected from an Evolution 2000 UV–Vis spectrometer (Thermo Fisher Scientific Inc., Madison, WI, USA) with $BaSO_4$ employed as the reference. The Brunauer–Emmett–Teller (BET) specific surface area of the Ag_3PO_4 microparticles was characterized using a PMI C-BET 201A system (Porous Materials Inc., Ithaca, NY, USA).

2.4. Photocatalytic Degradation Activity

The Ag_3PO_4 microparticles (S1–S4) were used to degrade MB and phenol under white-light LED irradiation (5 W, PCX-50C, Beijing perfectlight technology co. LTD, Beijing, China), and their photocatalytic activity was evaluated based on our earlier reports with modifications [33–37]. Prior to irradiation, 50 mL of 5 mg/L MB aqueous solution or 10 mg/L phenol solution containing 50 mg of photocatalyst was stirred in the dark for 30 min. During irradiation from the bottom with constant magnetic stirring, 3 mL of suspension was taken and centrifuged to remove the photocatalyst. The MB or phenol content in the filtrate was determined colorimetrically at 665 nm and 270 nm, respectively, by using a using a Synergy H1 Hybrid Multi-Mode Microplate Reader (Biotek Instruments, Inc., Winooski, VT, USA). Furthermore, the amount of total organic carbon was measured during degradation using an elementar Acquray TOC analyzer (Elementar Analysensysteme GmbH, Langenselbold, Germany). Similar procedures were performed for various dyestuffs (MR, RhB and acid blue 1) and environmental water samples (lake water and pond water), except for the use of solar irradiation.

2.5. Study of Recombination Rate of Electron–Hole Pairs and Charge Separation Efficiency

The recombination rate of the electron–hole pairs and charge separation efficiency for the Ag_3PO_4 microparticles (S1–S4) were evaluated based on our earlier reports [33–37]. For determination of recombination rate of electron–hole pairs, photoluminescence spectra were produced using a Varian Cary Eclipse fluorescence spectrometer (Agilent Technologies, Inc., Santa Clara, CA, USA). A slurry containing a photocatalyst (5 mg) and methanol (5 mL) was coated on the indium tin oxide (ITO) glass. The coated ITO glass was then heated on the hotplate at 50 °C for 30 min to completely remove the methanol. This coated ITO glass was used as an anode electrode, a platinum foil was used as a cathode electrode, and 0.1 M aqueous Na_2SO_4 solution was used as a supporting electrolyte. Photocurrents, which used to evaluate the charge separation efficiency, were measured using a CHI-6122E electrochemical analyzer (CH instruments, Inc., Austin, TX, USA) at room temperature.

2.6. Scavenger Test

The scavenger test was used to determine the major reactive species involved in the MB degradation. The procedure was similar to that the photocatalytic activity procedure [33–37]. Prior to adding a photocatalyst, hole-, oxygen radical- and hydroxyl radical- scavengers (each 10.0 µM) were added to the MB solution.

3. Results and Discussion

3.1. Crystalline Phase and Morphology

Different concentration effect on silver nitrate was examined by XRD and SEM to reveal their influence on crystalline phase and morphology of the Ag_3PO_4 microparticles. The XRD patterns of the various Ag_3PO_4 microparticles prepared from varying $AgNO_3$ concentration are presented in Figure 1.

The main diffraction peaks for all the types of Ag_3PO_4 microparticle can be indexed to a body-centered cubic structure (Joint Committee on Powder Diffraction Standards 06-0505), indicating that the $AgNO_3$ concentration have no influence on the crystalline phase of the Ag_3PO_4 microparticles.

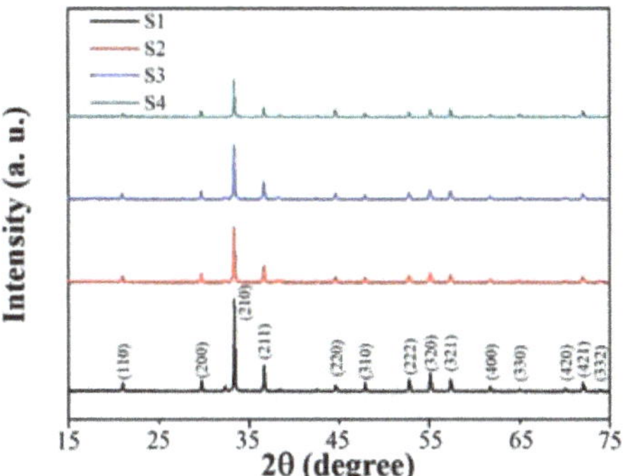

Figure 1. XRD patterns of the materials synthesized from different AgNO$_3$ concentrations (S1: 1.25 mmol, S2: 7.5 mmol, S3: 12.5 mmol, and S4: 25.0 mmol of AgNO$_3$.) under sonochemical process (600 W) at 25 kHz for 30 min with the hydrothermal process at 180 °C for 24 h.

Figure 2 shows SEM images of the Ag$_3$PO$_4$ microparticles, which reveal clear differences in the morphology and size of particles. The Ag$_3$PO$_4$ microparticle (S1), which was prepared by mixing 1.25 mmol of AgNO$_3$ with 2.5 mmol of Na$_3$PO$_4$ in 40.0 mL DI-water through the sonochemical and hydrothermal process, was constructed from spherical particles with a diameter of 4.5 ± 1.7 μm (Figure 1a). As the AgNO$_3$ content was increased gradually (7.50, 12.5 and then 25.0 mmol), the Ag$_3$PO$_4$ microparticles became irregular in shape and increased in size (Figure 1b: 7.3 ± 4.3 μm; Figure 1c: 11.3 ± 2.9 μm; and Figure 1d: 13.6 ± 5.0 μm). Because AgNO$_3$ was the precursor, a high concentration of AgNO$_3$ enhanced the growth rate of the Ag$_3$PO$_4$ particles; thus, the morphology changed from spherical to irregular, dependent on the AgNO$_3$ content.

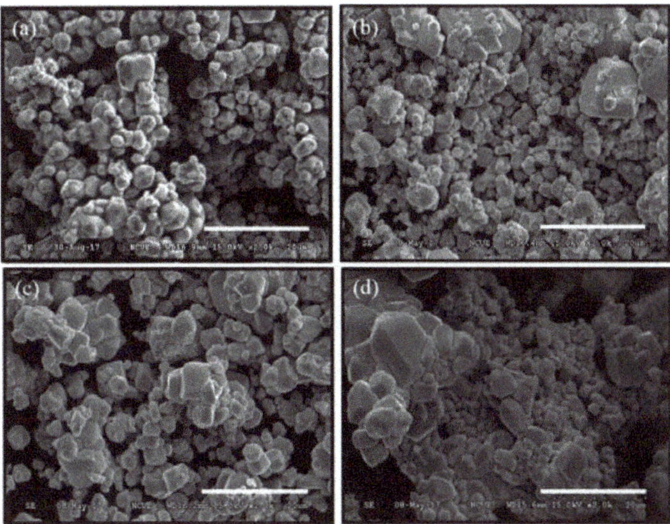

Figure 2. SEM images of the Ag$_3$PO$_4$ microparticles synthesized from different AgNO$_3$ concentrations under sonochemical (600 W) at 25 kHz for 30 min and hydrothermal process at 180 °C for 24 h: (**a**) 1.25 mmoL (S1), (**b**) 7.5 mmol (S2), (**c**) 12.5 mmol (S3), and (**d**) 25.0 mmol (S4) of AgNO$_3$. Scale bar: 20 μm.

3.2. Photophysical Properties and Specific Surface Area

Figure 3A displays the UV-Vis-DRS spectra of the Ag_3PO_4 microparticles, which clearly reveal that the yellow Ag_3PO_4 microparticles absorbed from UV to visible light range. The square root of the absorption coefficient was linearly correlated with energy, signifying that the absorption edges of all the Ag_3PO_4 microparticles were due to indirect transitions (Figure 3B). The energy bandgaps (Eg) of the microparticles were determined according to Tauc Plots [38]. Given that n for Ag_3PO_4 is 4 because of indirect transitions, the corresponding E_g of the Ag_3PO_4 microparticles were found from the x-axis intercept of the curve tangent in the plot of $(\alpha h\nu)^{0.5}$ versus hν (Figure 3B) [37]. The estimated band gaps of the Ag_3PO_4 microparticles were approximately in the range 2.32–2.41 eV (Table 1) [27]. The difference in band gap of microspheres generated from various $AgNO_3$ concentration may have caused different charge separation efficiency and recombination rate of the photogenerated electron–hole pairs, influencing photocatalytic activities of microspheres. The conduction band (CB) and valence band (VB) positions in the Ag_3PO_4 microparticles were determined according to the following equation: $E_{VB} = 1.46 + 0.5E_g$, where Eg is the band gap [27,38]. Through this equation, E_{VB} and E_{CB} for the Ag_3PO_4 microparticles were calculated (Table 1).

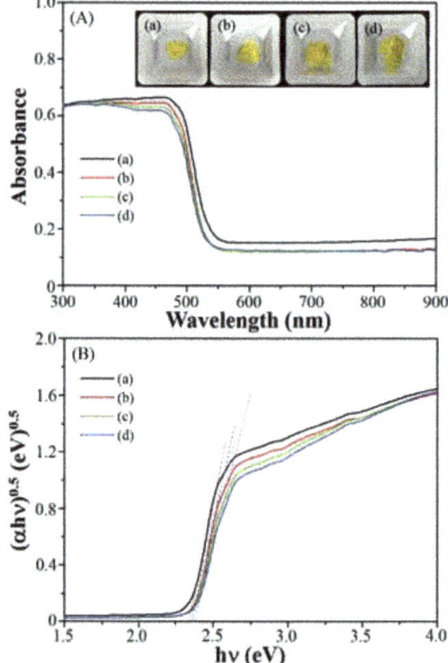

Figure 3. (**A**) UV–visible diffuse reflectance spectra (UV-Vis-DRS) and (**B**) $(\alpha h\nu)^{0.5}$ versus hν curves of Ag_3PO_4 (a) S1, (b) S2, (c) S3, and (d) S4. Inset: photographs of the corresponding materials.

Table 1. Measured band gap (Eg), conduction-band (E_{CB}) edge, valence-band (E_{VB}) edge, and specific surface area (S_{BET}) for the Ag_3PO_4 catalysts.

Series	Eg (eV)	E_{CB} (eV)	E_{VB} (eV)	S_{BET} (m$^2 \cdot$g^{-1})
S1	2.36	0.28	2.64	0.65
S2	2.38	0.27	2.65	0.55
S3	2.39	0.27	2.66	0.48
S4	2.41	0.26	2.67	0.36

The BET specific surface area of the Ag_3PO_4 microparticles was determined using nitrogen adsorption–desorption measurements; the corresponding values are listed in Table 1. Apparently, BET-determined surface area decreased with an increase in $AgNO_3$ concentration. This result is consistent with the size of the Ag_3PO_4 microparticles from SEM images. It is noteworthy to learn the difference in surface area may affect the extent of interaction between the photocatalyst and an organic pollutant, which may also cause different photocatalytic level.

3.3. Photocatalytic Degradation, Mechanism and Applications

The photocatalytic efficiency of the Ag_3PO_4 microparticles during MB degradation under white-light LED irradiation (5 W) was evaluated. Figure 4A plots the MB concentration ratio (C/C_0, where C_0 represents initial concentration and C represents concentration at time t) with irradiation time when different photocatalyst sample, S1–S4, was used. To make sure the Ag_3PO_4 microparticles were photocatalytically active, an experiment of direct MB photolysis in the absence of photocatalyst was performed under the same conditions. The MB concentration barely changed in the control experiment as the irradiation time increased (cyan curve). Under white-light LED irradiation for 10 min, S1 was clearly the most photocatalytically active (>99% degradation) among all other samples, S2–S4. Interestingly, Ag_3PO_4 microparticles prepared without hydrothermal process also exhibited >99% degradation within 10 min. However, they were not stable during the photocatalytic process (discussion later) because of photocorrosion [31,32,39]. In this study, S1 was the optimal candidate for MB degradation. According to the data in Figure 4A, the MB degradation was fitted as a pseudo-first-order reaction (Figure 4B). The rate constant and the half-life for MB degradation by the Ag_3PO_4 microparticles were determined (the corresponding values are listed in Table 2). Notably, S1 resulted in the highest rate constant and the shortest half-life for MB degradation. This may be because S1 possessed a large specific surface area, which resulted in high photodegradation activity.

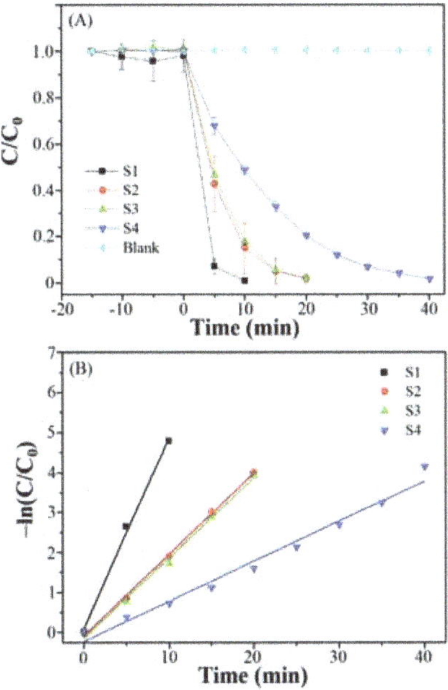

Figure 4. (**A**) Reduction of the concentration of methylene blue (MB) (C/C_0) with time and (**B**) corresponding rate constant of MB degradation by Ag_3PO_4 (S1–S4) under white-light LED irradiation (5 W).

Table 2. Pseudo-first-order rate constants and half-life for photocatalytic decoloration of MB by various Ag$_3$PO$_4$ photocatalysts.

Series	First-Order Kinetic Equation	k (min^{-1})	R^2	Half-Life (min)
S1	y = 0.478x + 0.100	0.478	0.9932	1.45
S2	y = 0.203x − 0.087	0.203	0.9978	3.41
S3	y = 0.200x − 0.147	0.200	0.9931	3.47
S4	y = 0.100x − 0.224	0.100	0.9783	6.93

Figure 5A shows the temporal evolution of the spectrum with MB degradation by S1. According to the literature, chromophore cleavage is analogous to the competitive photodegradation reaction involved in the organic pollutants decomposition [37]. In this study, the MB absorption at 665 nm decreased with an increase in the irradiation time. Moreover, a slight hypochromic shift was found in the characteristic absorption of MB in the presence of S1. Thus, the cleavage of the MB chromophore appears to predominate in Ag$_3$PO$_4$-photocatalytic decomposition systems, supported by the results from Figure 5B, which plots the temporal evolution of the total organic carbon content with MB degradation over S1. Although the colour of the MB solution disappeared under white-light LED irradiation for 10 min, the total organic carbon content only decreased to 1.39 mg/L (approximately decomposition 54.7% of original concentration). The photocatalytic decomposition of MB by S1 was increased as high as 71.0% (0.89 mg/L) under white-light LED irradiation for 120 min.

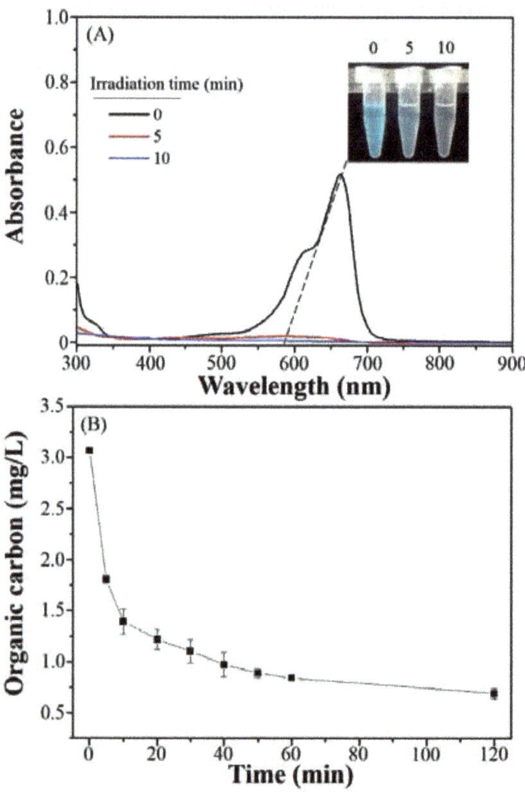

Figure 5. (**A**) Temporal evolution of the spectrum and (**B**) decomposition of MB by Ag$_3$PO$_4$ (S1) under white-light light-emitting-diode (LED) irradiation (5 W).

In order to exclude the MB sensitization under white-light LED irradiation, the photocatalytic activity of S1 was also evaluated by degradation of colorless phenol. The change in phenol concentration with irradiation time when S1 was used as shown in Figure 6. It is found that the concentration of phenol decreases with the irradiation time. The rate constant for degradation of phenol by S1 was 0.540 min^{-1}.

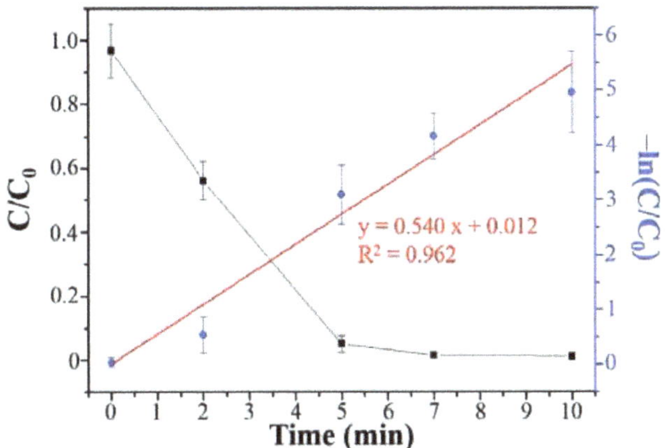

Figure 6. Reduction of the concentration of phenol (C/C$_0$) with time and the corresponding rate constant of phenol degradation by Ag$_3$PO$_4$ (S1) under white-light LED irradiation (5 W).

Besides the specific surface area of microparticles, the recombination rate of photogenerated electron–hole pairs and charge separation efficiency can affect the photocatalytic activity. Recombination of electron–hole pairs can release energy in the form of luminescence; thus, a low recombination rate suggests a lower luminescence intensity in the photoluminescence measurement and can correlate to higher photocatalytic activity of microspheres. Therefore, the luminescence spectra of the Ag$_3$PO$_4$ microparticles were measured at an excitation wavelength of 365 nm (Figure 7A). The luminescence intensity of S1 was the lowest among all samples (S1–S4), implying that the electron–hole pairs in S1 recombined the slowest. This may be because the lowest AgNO$_3$ concentration made Ag atoms on the Ag$_3$PO$_4$ microparticles surface coordinatively unsaturated (three-coordinated sites with one dangling bond) [40], inhibiting the recombination of photogenerated electrons and holes, and thus enhanced photocatalytic activity. The charge separation efficiency was evaluated through photocurrent measurements. Generally, a larger photocurrent implies higher separation efficiency and higher photocatalytic activity. The photocurrent of S1 was higher than those of other samples (Figure 7B), indicating enhanced separation of photogenerated electrons and holes. Therefore, the high photocatalytic activity of S1 was due to its large specific surface area, the low recombination rate of the photogenerated electron–hole pairs and the higher charge separation efficiency.

To resolve the mechanism responsible for the degradation of MB, radical- and hole-scavenger tests were implemented to identify the main active species in the photocatalytic degradation process when S1 was used. Under white-light LED irradiation, MB degradation was deceased upon the addition of a hole scavenger (EDTA) and an oxygen radical scavenger (p-benzoquinone), as illustrated in Figure 8. This indicates that hole and oxygen radicals are the major reactive species involved in the MB decomposition.

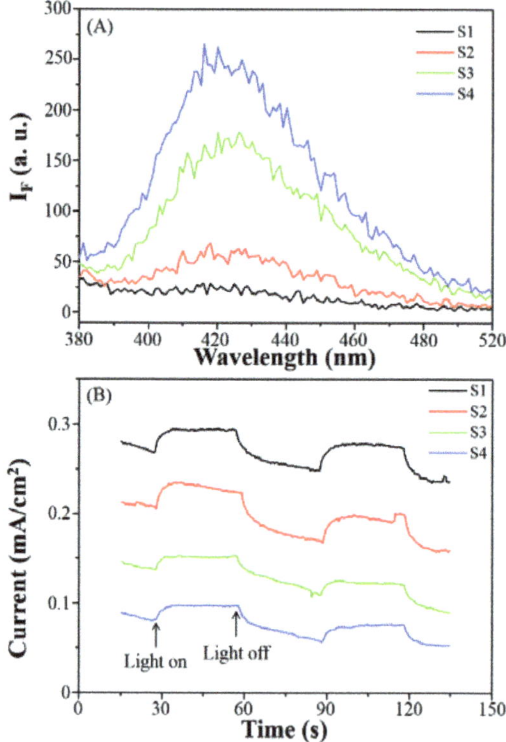

Figure 7. (**A**) Photoluminescence at λ_{ex} = 365 nm and (**B**) photocurrents spectra measured of Ag_3PO_4 (S1–S4).

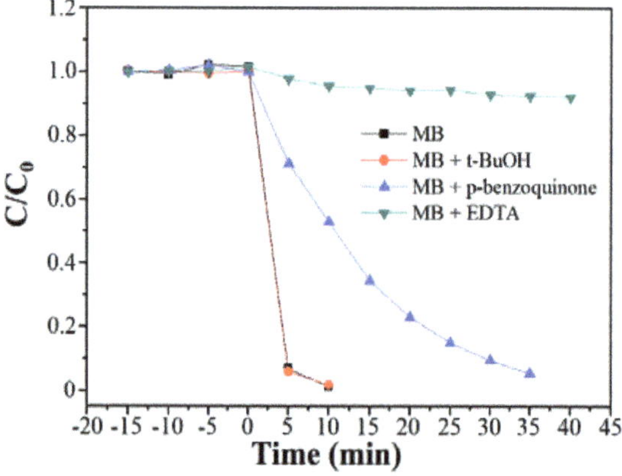

Figure 8. Plots of photogenerated carrier trapping for MB decolorization catalyzed by Ag_3PO_4 (S1).

A possible degradation mechanism is proposed in Scheme 1. When Ag_3PO_4 was illuminated under white-light LED irradiation, electrons in the VB were promoted to the CB, with an equal number of holes produced in the VB (Equation (1)). Then, the electrons in the CB were transferred to the (111) surface of the Ag_3PO_4 microparticles, which decreased the electron–hole recombination rate.

The electrons on the (111) surface were unstable and could react with oxygen to produce oxygen radicals (Equations (2) and (3)) [17,20,22–25]. Therefore, holes in the VB and oxygen radicals were the major reactive species in the degradation of MB (Equations (4) and (5)).

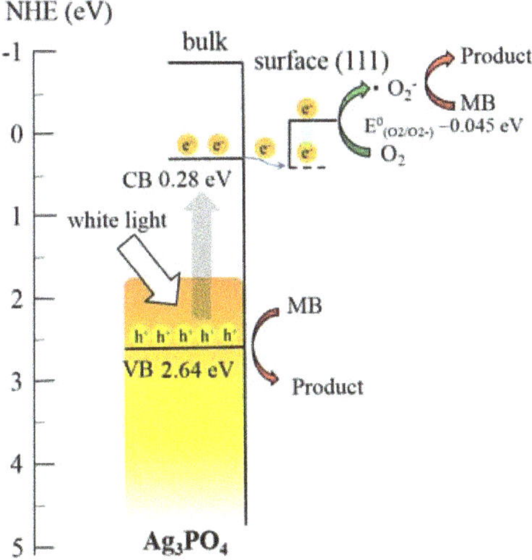

Scheme 1. Illustration of the photocatalytic degradation mechanism of the Ag$_3$PO$_4$ microparticles under white-light LED irradiation.

$$Ag_3PO_4 + h\nu \rightarrow Ag_3PO_4\ (h^+ + e^-) \tag{1}$$

$$Ag_3PO_4\ (h^+ + e^-) \rightarrow Ag_3PO_4\ (h^+{}_{VB}) + Ag_3PO_4\ (e^-{}_{CB}) \tag{2}$$

$$Ag_3PO_4\ (e^-{}_{CB}) + O_2 \rightarrow {}^{\bullet}O_2{}^- \tag{3}$$

$${}^{\bullet}O_2{}^- + MB \rightarrow \text{oxidized products} \tag{4}$$

$$Ag_3PO_4\ (h^+{}_{VB}) + MB \rightarrow \text{oxidized products} \tag{5}$$

In addition to the photocatalytic efficiency of the photocatalyst, the stability of Ag$_3$PO$_4$ (S1) is vital for practical applications. To assess the stability and efficiency of the Ag$_3$PO$_4$ microparticles, a prolonged run of MB degradation was performed. After three cycles for 10 min in each cycle, a decrease in the photodegradation rate was observed when the Ag$_3$PO$_4$ microparticles prepared without hydrothermal process was employed (black curve in Figure 9A) while the MB photodegradation activity of S1 remained constant even after five cycles (red curve in Figure 9A), maintaining 98.7% of activity within 10 min after five cycles. This indicated that S1 had an excellent stability, which was proved by the lack of change in the XRD pattern after five photolysis cycles (Figure 9B). This may be because the hydrothermal procedure enhanced the degree of crystallinity in the microparticles [29].

To assess the practical applications of S1, various dyestuffs (MB, MR, RhB and acid blue 1) in environmental water samples (lake water and pond water) were degraded under sunlight (Figure 10). S1 exhibited excellent photocatalytic activity under sunlight for the degradation of all dyestuffs, with nearly 90% degradation within 50 min. A remarkable difference in the degradation time for all dyestuffs was observed for the environmental water samples (90% degradation within 50 min) compared with the deionized water sample (100% degradation within 10 min). This may be because the presence of anions or radical scavengers in the environmental water samples reduced the photocatalytic activity of S1. Further research on the enhancing photocatalytic activity of Ag$_3$PO$_4$ composites, such as those wrapped by reduced graphene oxide and g-C$_3$N$_4$, is underway in our laboratory.

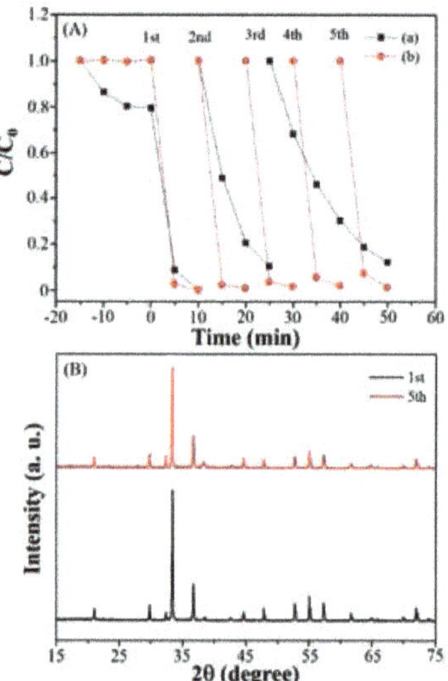

Figure 9. (**A**) Cycling runs of MB photocatalytic decolorization in the presence of the Ag$_3$PO$_4$ microparticles prepared (a) without and (b) with hydrothermal process. (**B**) XRD patterns for S1 before and after being used for five reaction cycles.

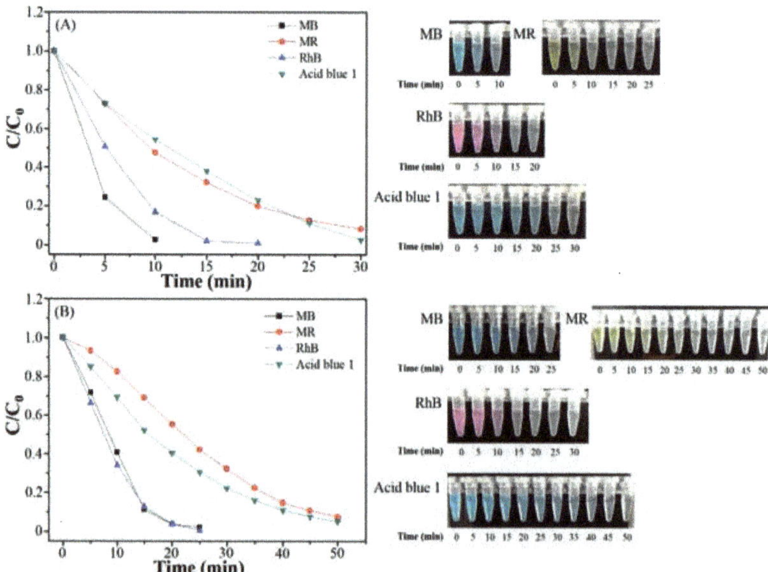

Figure 10. Decoloration kinetic curves and photographs of different organic dyes (MB, MR, RhB and acid blue 1, each 10 ppm), degraded using S1 in (**A**) lake water and (**B**) pond water under solar light irradiation.

4. Conclusions

The Ag_3PO_4 microparticles were synthesized using a combination of sonochemical and hydrothermal processes from various concentrations of $AgNO_3$. The photocatalytic activity of the Ag_3PO_4 microparticles was evaluated through the degradation of MB under low-power white-light LED irradiation. The results reveal that S1 synthesized from 1.25 mmol $AgNO_3$ in 40.0 mL DI-water displayed remarkably higher photocatalytic degradation activity (>99.8%) within 10 min than other microparticle types (S2–S4). This can be attributed to its large specific surface area, the low recombination rate of its photogenerated electron–hole pairs and the higher charge separation efficiency. This study shows that the strong photocatalytic activity of the Ag_3PO_4 microparticles (S1) was considered to be possible for its applications toward photodegradation of organic pollutes in environmental water samples.

Author Contributions: Y.-W.L. conceived and designed the experiments; C.-S.T. performed the experiments; C.-S.T. and T.W. analyzed the data; Y.-W.L contributed reagents/materials/analysis tools and wrote the paper. T.W. contributed the English language editing.

Acknowledgments: This study was supported by the Ministry of Science and Technology under contract (MOST 106-2119-M-018-001).

Conflicts of Interest: The authors declare no conflict of interest.

References

1. Ince, N.H. Ultrasound-Assisted Advanced Oxidation Processes for Water Decontamination. *Ultrason. Sonochem.* **2018**, *40*, 97–103. [CrossRef] [PubMed]
2. Ayed, L.; Asses, N.; Chammem, N.; Ben Othman, N.; Hamdi, M. Advanced Oxidation Process and Biological Treatments for Table Olive Processing Wastewaters: Constraints and a Novel Approach to Integrated Recycling Process: A Review. *Biodegradation* **2017**, *28*, 125–138. [CrossRef] [PubMed]
3. Giannakis, S.; Rtimi, S.; Pulgarin, C. Light-Assisted Advanced Oxidation Processes for the Elimination of Chemical and Microbiological Pollution of Wastewaters in Developed and Developing Countries. *Molecules* **2017**, *22*, 1070. [CrossRef] [PubMed]
4. Trojanowicz, M.; Bojanowska-Czajka, A.; Capodaglio, A.G. Can Radiation Chemistry Supply a Highly Efficient Ao(R)P Process for Organics Removal from Drinking and Waste Water? A Review. *Environ. Sci. Pollut. Res.* **2017**, *24*, 20187–20208. [CrossRef] [PubMed]
5. Alvarez-Corena, J.R.; Bergendahl, J.A.; Hart, F.L. Advanced Oxidation of Five Contaminants in Water by UV/TiO$_2$: Reaction Kinetics and Byproducts Identification. *J. Environ. Manag.* **2016**, *181*, 544–551. [CrossRef] [PubMed]
6. Bethi, B.; Sonawane, S.H.; Rohit, G.S.; Holkar, C.R.; Pinjari, D.V.; Bhanvase, B.A.; Pandit, A.B. Investigation of TiO$_2$ Photocatalyst Performance for Decolorization in the Presence of Hydrodynamic Cavitation as Hybrid Aop. *Ultrason. Sonochem.* **2016**, *28*, 150–160. [CrossRef] [PubMed]
7. Havlikova, L.; Satinsky, D.; Solich, P. Aspects of Decontamination of Ivermectin and Praziquantel from Environmental Waters Using Advanced Oxidation Technology. *Chemosphere* **2016**, *144*, 21–28. [CrossRef] [PubMed]
8. Tokumura, M.; Sugawara, A.; Raknuzzaman, M.; Habibullah-Al-Mamun, M.; Masunaga, S. Comprehensive Study on Effects of Water Matrices on Removal of Pharmaceuticals by Three Different Kinds of Advanced Oxidation Processes. *Chemosphere* **2016**, *159*, 317–325. [CrossRef] [PubMed]
9. Villegas-Guzman, P.; Silva-Agredo, J.; Florez, O.; Giraldo-Aguirre, A.L.; Pulgarin, C.; Torres-Palma, R.A. Selecting the Best Aop for Isoxazolyl Penicillins Degradation as a Function of Water Characteristics: Effects of Ph, Chemical Nature of Additives and Pollutant Concentration. *J. Environ. Manag.* **2017**, *190*, 72–79. [CrossRef] [PubMed]
10. Alvarado-Morales, M.; Tsapekos, P.; Awais, M.; Gulfraz, M.; Angelidaki, I. TiO$_2$/UV Based Photocatalytic Pretreatment of Wheat Straw for Biogas Production. *Anaerobe* **2017**, *46*, 155–161. [CrossRef] [PubMed]
11. Bohdziewicz, J.; Kudlek, E.; Dudziak, M. Influence of the Catalyst Type (TiO$_2$ and Zno) on the Photocatalytic Oxidation of Pharmaceuticals in the Aquatic Environment. *Desalin. Water Treat.* **2016**, *57*, 1552–1563. [CrossRef]

12. Dong, H.R.; Zeng, G.M.; Tang, L.; Fan, C.Z.; Zhang, C.; He, X.X.; He, Y. An Overview on Limitations of TiO$_2$-Based Particles for Photocatalytic Degradation of Organic Pollutants and the Corresponding Countermeasures. *Water Res.* **2015**, *79*, 128–146. [CrossRef] [PubMed]

13. Liu, H.; Liu, S.; Zhang, Z.; Dong, X.; Liu, T. Hydrothermal Etching Fabrication of TiO$_2$@Graphene Hollow Structures: Mutually Independent Exposed {001} and {101} Facets Nanocrystals and Its Synergistic Photocaltalytic Effects. *Sci. Rep.* **2016**, *6*, 33839. [CrossRef] [PubMed]

14. Veisi, F.; Zazouli, M.A.; Ebrahimzadeh, M.A.; Charati, J.Y.; Dezfoli, A.S. Photocatalytic Degradation of Furfural in Aqueous Solution by N-Doped Titanium Dioxide Nanoparticles. *Environ. Sci. Pollut. Res.* **2016**, *23*, 21846–21860. [CrossRef] [PubMed]

15. Reddy, K.R.; Hassan, M.; Gomes, V.G. Hybrid Nanostructures Based on Titanium Dioxide for Enhanced Photocatalysis. *Appl. Catal. A Gen.* **2015**, *489*, 1–16. [CrossRef]

16. Yi, Z.; Ye, J.; Kikugawa, N.; Kako, T.; Ouyang, S.; Stuart-Williams, H.; Yang, H.; Cao, J.; Luo, W.; Li, Z.; et al. An Orthophosphate Semiconductor with Photooxidation Properties under Visible-Light Irradiation. *Nat. Mater.* **2010**, *9*, 559–564. [CrossRef] [PubMed]

17. Yan, X.; Gao, Q.; Qin, J.; Yang, X.; Li, Y.; Tang, H. Morphology-Controlled Synthesis of Ag$_3$PO$_4$ Microcubes with Enhanced Visible-Light-Driven Photocatalytic Activity. *Ceram. Int.* **2013**, *39*, 9715–9720. [CrossRef]

18. Yang, Z.-M.; Liu, Y.-Y.; Xu, L.; Huang, G.-F.; Huang, W.-Q. Facile Shape-Controllable Synthesis of Ag$_3$PO$_4$ Photocatalysts. *Mater. Lett.* **2014**, *133*, 139–142. [CrossRef]

19. Dong, L.; Wang, P.; Wang, S.; Lei, P.; Wang, Y. A Simple Way for Ag$_3$PO$_4$ Tetrahedron and Tetrapod Microcrystals with High Visible-Light-Responsive Activity. *Mater. Lett.* **2014**, *134*, 158–161. [CrossRef]

20. Hao, Z.; Ai, L.; Zhang, C.; Niu, Z.; Jiang, J. Self-Sacrificial Synthesis of Porous Ag$_3$PO$_4$ Architectures with Enhanced Photocatalytic Activity. *Mater. Lett.* **2015**, *143*, 51–54. [CrossRef]

21. Dong, P.; Yin, Y.; Xu, N.; Guan, R.; Hou, G.; Wang, Y. Facile Synthesis of Tetrahedral Ag$_3$PO$_4$ Mesocrystals and Its Enhanced Photocatalytic Activity. *Mater. Res. Bull.* **2014**, *60*, 682–689. [CrossRef]

22. Wan, J.; Sun, L.; Fan, J.; Liu, E.; Hu, X.; Tang, C.; Yin, Y. Facile Synthesis of Porous Ag$_3$PO$_4$ Nanotubes for Enhanced Photocatalytic Activity under Visible Light. *Appl. Surf. Sci.* **2015**, *355*, 615–622. [CrossRef]

23. Guo, X.; Chen, C.; Yin, S.; Huang, L.; Qin, W. Controlled Synthesis and Photocatalytic Properties of Ag$_3$PO$_4$ Microcrystals. *J. Alloys Compd.* **2015**, *619*, 293–297. [CrossRef]

24. Li, J.; Ji, X.; Li, X.; Hu, X.; Sun, Y.; Ma, J.; Qiao, G. Preparation and Photocatalytic Degradation Performance of Ag$_3$PO$_4$ with a Two-Step Approach. *Appl. Surf. Sci.* **2016**, *372*, 30–35. [CrossRef]

25. Xie, Y.; Huang, Z.; Zhang, Z.; Zhang, X.; Wen, R.; Liu, Y.; Fang, M.; Wu, X. Controlled Synthesis and Photocatalytic Properties of Rhombic Dodecahedral Ag$_3$PO$_4$ with High Surface Energy. *Appl. Surf. Sci.* **2016**, *389*, 56–66. [CrossRef]

26. Frontistis, Z.; Antonopoulou, M.; Petala, A.; Venieri, D.; Konstantinou, I.; Kondarides, D.I.; Mantzavinos, D. Photodegradation of Ethyl Paraben Using Simulated Solar Radiation and Ag$_3$PO$_4$ Photocatalyst. *J. Hazard. Mater.* **2017**, *323*, 478–488. [CrossRef] [PubMed]

27. Hsieh, M.-S.; Su, H.-J.; Hsieh, P.-L.; Chiang, Y.-W.; Huang, M.-H. Synthesis of Ag$_3$PO$_4$ Crystals with Tunable Shapes for Facet-Dependent Optical Property, Photocatalytic Activity, and Electrical Conductivity Examinations. *ACS Appl. Mater. Interfaces* **2017**, *9*, 39086–39093. [CrossRef] [PubMed]

28. Cui, X.; Zheng, Y.F.; Zhou, H.; Yin, H.Y.; Song, X.C. The Effect of Synthesis Temperature on the Morphologies and Visible Light Photocatalytic Performance of Ag$_3$PO$_4$. *J. Taiwan Inst. Chem. Eng.* **2016**, *60*, 328–334. [CrossRef]

29. Zhang, S.; Gu, X.; Zhao, Y.; Qiang, Y. Effect of Annealing Temperature and Time on Structure, Morphology and Visible-Light Photocatalytic Activities Ag$_3$PO$_4$ Microparticles. *Mater. Sci. Eng. B Solid State Mater. Adv. Technol.* **2015**, *201*, 57–65. [CrossRef]

30. Huang, G.-F.; Ma, Z.-L.; Huang, W.-Q.; Tian, Y.; Jiao, C.; Yang, Z.-M.; Wan, Z.; Pan, A. Ag$_3$PO$_4$ semiconductor Photocatalyst: Possibilities and Challenges. *J. Nanomater.* **2013**, *2013*, 1–8.

31. Liu, J.H.; Li, X.; Liu, F.; Lu, L.H.; Xu, L.; Liu, L.W.; Chen, W.; Duan, L.M.; Liu, Z.R. The Stabilization Effect of Surface Capping on Photocatalytic Activity and Recyclable Stability of Ag$_3$PO$_4$. *Catal. Commun.* **2014**, *46*, 138–141. [CrossRef]

32. Luo, L.; Li, Y.Z.; Hou, J.T.; Yang, Y. Visible Photocatalysis and Photostability of Ag$_3$PO$_4$ Photocatalyst. *Appl. Surf. Sci.* **2014**, *319*, 332–338. [CrossRef]

33. Chen, Y.-J.; Tseng, C.-S.; Tseng, P.-J.; Huang, C.-W.; Wu, T.; Lin, Y.-W. Synthesis and Characterization of Ag/Ag$_3$PO$_4$ Nanomaterial Modified BiPO$_4$ Photocatalyst by Sonochemical Method and Its Photocatalytic Application. *J. Mater. Sci. Mater. Electron.* **2017**, *28*, 11886–11899. [CrossRef]

34. Cheng, L.-W.; Tsai, J.-C.; Huang, T.-Y.; Huang, C.-W.; Unnikrishnan, B.; Lin, Y.-W. Controlled Synthesis, Characterization and Photocatalytic Activity of BiPO$_4$ Nanostructures with Different Morphologies. *Mater. Res. Express* **2014**, *1*. [CrossRef]

35. Huang, C.-K.; Wu, T.; Huang, C.-W.; Lai, C.-Y.; Wu, M.-Y.; Lin, Y.-W. Enhanced Photocatalytic Performance of BiVO$_4$ in Aqueous AgNO$_3$ Solution under Visible Light Irradiation. *Appl. Surf. Sci.* **2017**, *399*, 10–19. [CrossRef]

36. Huang, C.-W.; Wu, M.-Y.; Lin, Y.-W. Solvothermal Synthesis of Ag Hybrid BiPO$_4$ Heterostructures with Enhanced Photodegradation Activity and Stability. *J. Colloid Interf. Sci.* **2017**, *490*, 217–225. [CrossRef] [PubMed]

37. Huang, T.-Y.; Chen, Y.-J.; Lai, C.-Y.; Lin, Y.-W. Synthesis, Characterization, Enhanced Sunlight Photocatalytic Properties, and Stability of Ag/Ag$_3$PO$_4$ Nanostructure-Sensitized BiPO$_4$. *RSC Adv.* **2015**, *5*, 43854–43862. [CrossRef]

38. Lin, H.L.; Ye, H.F.; Xu, B.Y.; Cao, J.; Chen, S.F. Ag$_3$PO$_4$ Quantum Dot Sensitized BiPO$_4$: A Novel P-N Junction Ag$_3$PO$_4$/BiPO$_4$ with Enhanced Visible-Light Photocatalytic Activity. *Catal. Commun.* **2013**, *37*, 55–59. [CrossRef]

39. Ma, X.L.; Li, H.H.; Wang, Y.H.; Li, H.; Liu, B.; Yin, S.; Sato, T. Substantial Change in Phenomenon of "Self-Corrosion" on Ag$_3$PO$_4$/TiO$_2$ Compound Photocatalyst. *Appl. Catal. B Environ.* **2014**, *158*, 314–320. [CrossRef]

40. Wang, H.; Bai, Y.S.; Yang, J.T.; Lang, X.F.; Li, J.H.; Guo, L. A Facile Way to Rejuvenate Ag$_3$PO$_4$ as a Recyclable Highly Efficient Photocatalyst. *Chem. Eur. J.* **2012**, *18*, 5524–5529. [CrossRef] [PubMed]

© 2018 by the authors. Licensee MDPI, Basel, Switzerland. This article is an open access article distributed under the terms and conditions of the Creative Commons Attribution (CC BY) license (http://creativecommons.org/licenses/by/4.0/).

 *materials*

Article

Synthesis and Broadband Spectra Photocatalytic Properties of $Bi_2O_2(CO_3)_{1-x}S_x$

Junping Ding [1,2], Huanchun Wang [1,3], Haomin Xu [1], Lina Qiao [1], Yidong Luo [1], Yuanhua Lin [1,*] and Cewen Nan [1]

[1] State Key Laboratory of New Ceramics and Fine Processing, School of Materials Science and Engineering, Tsinghua University, Beijing 100084, China; djp15@mails.tsinghua.edu.cn (J.D.); huanchwang@163.com (H.W.); xuhm13@mails.tsinghua.edu.cn (H.X.); qln13@mails.tsinghua.edu.cn (L.Q.); ydluozd@163.com (Y.L.); cwnan@mail.tsinghua.edu.cn (C.N.)
[2] China Astronaut Research and Training Center, Beijing 100094, China
[3] High-Tech Institute of Xi'an, Xi'an 710025, China
* Correspondence: linyh@mail.tsinghua.edu.cn; Tel.: +86-10-6277-3741

Received: 31 March 2018; Accepted: 3 May 2018; Published: 14 May 2018

Abstract: High efficiency photocatalyst $Bi_2O_2(CO_3)_{1-x}S_x$ was synthesized conveniently with chemical bath precipitation using $Bi_2O_2CO_3$ as the precursor. The microstructures of the samples are systematically characterized by X-ray diffraction (XRD), scanning electron microscopy (SEM), high resolution transmission electron microscopy (HRTEM), X-ray photoelectron spectroscopy (XPS), ultraviolet photoelectron spectroscopy (UPS) and UV-Vis spectroscopy; the optical and photocatalytic properties are carefully tested as well. The content of S, which was tuned through the controlling of the precipitation process, was verified to have an intense effect over the photocatalytic properties. A nearly saturated S ratio and the best photocatalytic performance were observed in specimens with the most S content. Our study reveals that, with negligible influence of the morphology and crystal structure, $Bi_2O_2(CO_3)_{1-x}S_x$ possessed a broadened optical absorption regionfromultraviolet to visible light, and enhanced photocatalytic activity in comparison to precursor $Bi_2O_2CO_3$ in photocatalytic degradation of Congo Red aqueous solution.

Keywords: $Bi_2O_2CO_3$; $Bi_2O_2(CO_3)_{1-x}S_x$; broadband spectra; photocatalysis

1. Introduction

Semiconductor photocatalysis has attracted increasing attention because of the capability of harvesting the solar energy to eliminate environmental pollutants [1–7]. Among various semiconductors, some Aurivillius type bismuth-based oxide semiconductor materials such as BiOX (X = Cl, Br, I), $BiVO_4$ and Bi_2WO_6 have been widely used in photocatalysis [8–14].

Bismuth-based layered-structure compounds have a unique crystal structure and band structure. Hybridisation between 6s electrons of Bi and 2p electrons of O form chemical bonds which are stronger than those between Bi and other nonmetallic atoms (such as chalcogen), leading to a particularly stable (Bi-O)$^+$ layer. A series of Bi-based layered-structural photocatalytic materials of various band gap widths from 3.2 eV (e.g., BiOCl [15]) to 1.12~1.5 eV (e.g., Bi_2O_2S [16,17]) can be obtained by combining the (Bi-O)$^+$ layer with different anion layers. In addition, p-type (BiCuSO or the like) or n-type ($Bi_2O_2CO_3$, etc.) semiconductor materials can be obtained by adjusting the anion layer. Therefore, the different Bi-based oxide composite structure can not only control and broaden the range of light absorption of the catalyst, but also may form a hetero structure such as p-n junction.

Recently, $Bi_2O_2CO_3$, which is a member of the Aurivillius-type family and composed of $[Bi_2O_2]^{2+}$ layers interleaved by CO_3^{2-} layers [18,19], has attracted growing concern because of its photocatalytic ability to decompose organic pollutants in liquid phase and NO in gaseous phase [20–22]. Its unique layered structure, resulting in a large internal electrostatic field and asymmetric polarization effect, contributes to the separation of photogenerated electron-hole pairs [23,24]. However, the application of $Bi_2O_2CO_3$ in photodegradation is strongly limited by its large band gap (~3.3 eV). To overcome this limitation, many methods have been developed, such as the fabrication of heterojunctions such as $BiVO_4/Bi_2O_2CO_3$, $Bi_2S_3/Bi_2O_3/Bi_2O_2CO_3$ [25–27], noble metal deposition [28], elemental doping [29], and morphological modulation [30].

In this paper, we have synthesized S-doped $Bi_2O_2(CO_3)_{1-x}S_x$ by chemical bath precipitation, using $Bi_2O_2CO_3$ as the precursor, through the controlling of the precipitation process to have an intense effect over the photocatalytic properties. A nearly saturated S ratio and the best photocatalytic performance were observed in specimens with the most S content. With a negligible influence of the morphology and crystal structure, the optical absorption of $Bi_2O_3CO_3$ was extended from the ultraviolet (UV) to the visible region. The photocatalytic degradation of Congo Red showed that $Bi_2O_2(CO_3)_{1-x}S_x$ exhibited enhanced photoactivity in comparison to the precursor powder.

2. Results and Discussion

2.1. Synthetic $Bi_2O_2(CO_3)_{1-x}S_x$

Figure 1 shows the XRD(X-ray diffraction) pattern of the $Bi_2O_2CO_3$ powder prepared by hydrothermal method, together with a reference pattern of tetragonal $Bi_2O_2CO_3$ (JCPDS: 41−1488). No second phase can be found, and the sharp peaks indicate well-developed crystallinity. The preparation process of $Bi_2O_2CO_3$ can be summarized in Equations (1)–(3). CO_3^{2-} forms through a hydrolysis reaction between $(NH_2)_2CO$ and H_2O. Bi_2O_3 is also strongly hydrolyzed with water to produce $(Bi_2O_2)^{2+}$. The produced $(Bi_2O_2)^{2+}$ and CO_3^{2-} then react to generate $Bi_2O_2CO_3$.

$$(NH_2)_2CO + 2H_2O \rightarrow 2NH_3^+ + CO_3^{2-} \tag{1}$$

$$Bi_2O_3 + 2H_2O \rightarrow (Bi_2O_2)^{2+} + 2OH^+ \tag{2}$$

$$(Bi_2O_2)^{2+} + CO_3^{2-} \rightarrow Bi_2O_2CO_3 \tag{3}$$

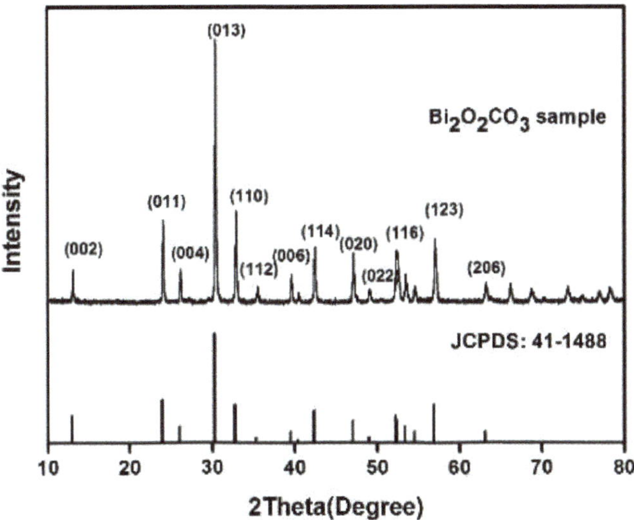

Figure 1. XRD pattern of hydrothermal synthesised $Bi_2O_2CO_3$.

In addition, the percentage of crystallinity and the BET (Brunauer–Emmett–Teller) specific surface area of the samples with a S:$Bi_2O_2CO_3$ ratio n equals to 0, 0.01, 0.02, 0.05, 0.10 and 0.20 (marked as M0, M1, M2, M5, M10 and M20, respectively) are shown in Table 1. There are no significant changes in their percentage of crystallinity, while samples of M5 and M10 displaylarger specific surface areas than that of other samples, which could lead to theexposure of more active sites for the photocatalytic experiment. The scanning electron microscopy (SEM) photograph and the high resolution transmission electron microscopy (HRTEM) images of the powder are shown in Figures 2 and 3. The morphology of the particles are nano-sized flakes of about 60–80 nm in thickness. In addition, the crystallinity of different samples calculated from the XRD results shows that *S* doping introduced defects in the $Bi_2O_2CO_3$ and thus caused crystallinity change.

Table 1. Surface area and percentage of crystallinity of the $Bi_2O_2CO_3$ and M1~M20 powders.

	M0	M1	M2	M5	M10	M20
Surface area (m^2/g)	0.917	0.973	0.980	1.666	1.823	0.966
Percentage of crystallinity (%)	–	74.43 ± 0.96	65.62 ± 0.61	74.14 ± 0.88	77.31 ± 0.75	79.87 ± 1.75

Figure 2. SEM photograph of hydrothermal synthesised $Bi_2O_2CO_3$.

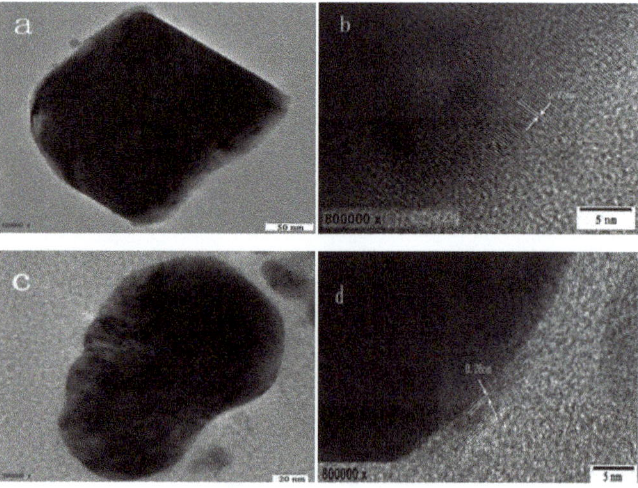

Figure 3. HRTEM images of M0 (**a**); (**b**) and M20 (**c**); (**d**).

The XRD patterns of the samples prepared by the Na_2S chemical bath precipitationare shown in Figure 4a. All diffraction peaks are consistent with $Bi_2O_2CO_3$, indicating that chemical bath precipitation did not introduce a significant second phase. The intensity of the diffraction peak does not obviously decrease, and the products still have good crystallinity. The position of the (013) diffraction peak for different samples are shown in Figure 4b. No obvious influence of Na_2S chemical precipitation on the crystal structure of $Bi_2O_2CO_3$ can be found because the position of the peak (013) did not show an apparent shift according to XRD results.

X-ray photoelectron spectroscopy (XPS) was utilized toobtain insights into the valence states and surface chemical compositions details of $Bi_2O_2(CO_3)_{1-x}S_x$. As shown in Figure 5a, the XPS spectrum of Bi-4f shows two peaks at 159.05 and 164.35 eV, which belong to Bi-4f$_{7/2}$ and Bi-4f$_{5/2}$ energy levels, respectively. These two peaks are characteristic features of trivalent Bi in $Bi_2O_2(CO_3)_{1-x}S_x$ [31]. The two peaks at 284.7 eV and 288.8 eV in Figure 5b show that the existence form of C is CO_3^{2-} [32]. In Figure 5c, the two peaks are at 530.5 eV and 531 eV, which belong to O energy levels in B-O and CO_3^{2-}, respectively [33]. In Figure 5d, the peak of S-2p is at the range of 158–166 eV, which shows that the existence form of S is S^{2-} [17]. On the other hand, the Bi-4f peak of M20 apparently shifts compared to M0 (Figure 5a), which proves that S takes place of CO_3^{2-} partially [34].

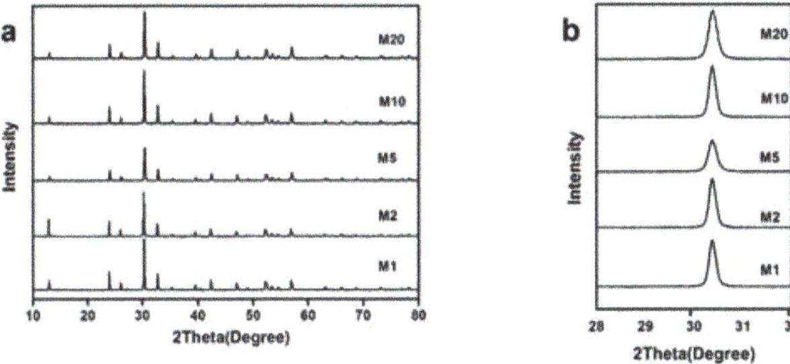

Figure 4. (**a**) XRD patterns of $Bi_2O_2(CO_3)_{1-x}S_x$ prepared by chemical bath precipitation; (**b**) the position of the (013) diffraction.

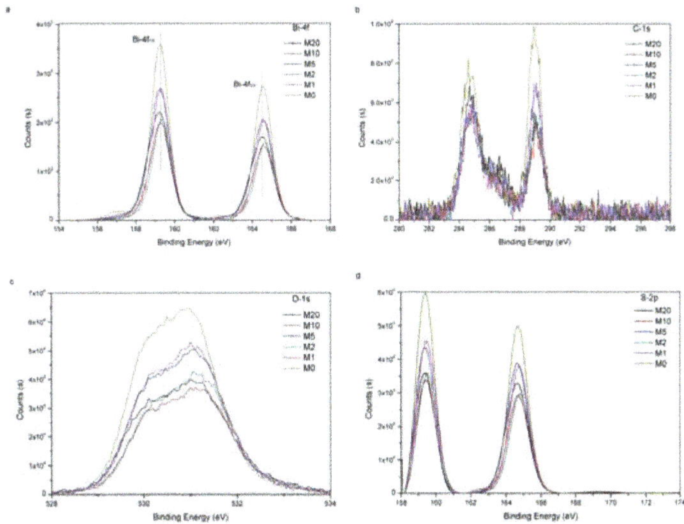

Figure 5. XPS spectra of M0, M1, M2, M5, M10 and M20, Bi-4f (**a**); C−1s (**b**); O−1s (**c**); and S-2p (**d**).

Although Na_2S chemical precipitation had no obvious influence on the crystal structure of $Bi_2O_2CO_3$, the powder color was changed from white to yellow, and the color became darker as S: $Bi_2O_2CO_3$ molar ratio n increased. The UV-Vis diffuse reflectance spectra are shown in Figure 6. $Bi_2O_2CO_3$ has a strong absorption of UV light with wavelengths less than 360 nm and weak absorption to 400 nm~500 nm-wavelength-visible light due to defects and oxygen vacancy, which also explained the fact that $Bi_2O_2CO_3$ could display visible light photocatalytic activity with the bandgap of 3.2 eV. With the introduction of S, the light absorption behaviour was significantly changed from M1 to M20. In particular, the absorption of visible light increased by about one order of magnitude. The band gap of $Bi_2O_2CO_3$ without sulfur is fitted as 3.27 eV, and the introduction of S leads to the emergenceof a narrow band gapby lowering the conduction band position and meanwhile generating impurity levels [35,36]. The adsorption edge is around 380 nm. With the increase of S content, defects and oxygen vacancies increase, possibly due to point defects, and the fitted narrow band gap decreases from 3.25 to 2.20 eV. Energy levels of the valence band maximum (E_{VB}) were measured by the ultraviolet photoelectron spectrometer at UV intensity 500 nW and energy levels of the conduction band minimum (E_{CB}) were calculated by the bandgap. As shown in Figure 7, valence band edge position and conduction band edge position become more negative after the incorporation of sulfur into $Bi_2O_2CO_3$.

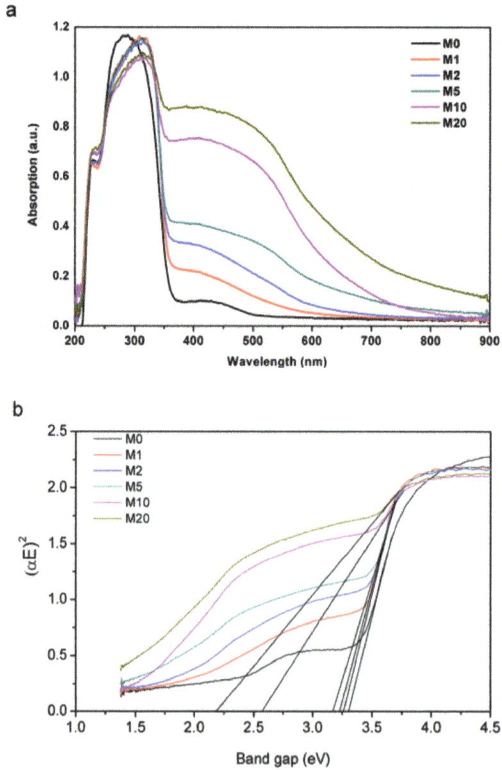

Figure 6. (**a**) UV-Vis diffuse reflectance spectra of the synthesized $Bi_2O_2(CO_3)_{1-x}S_x$; (**b**) band gap fitting with K-M relation.

The SEM observation showed that chemical bath treatment had little influence on the morphology of the $Bi_2O_2CO_3$ particles. EDS (energy dispersive spectroscopy) elemental mapping in Figure 8 revealed the homogeneous distribution of S on the particle surface, and no obvious segregation and aggregation can be seen among the particles. The quantitative elemental analysis results are shown in Table 2. The S content in samples M1~M20 increases with the increase of S: $Bi_2O_2CO_3$ molar ratio n, but the S atom percentage (referring to Bi-content) is obviously smaller than n and becomes stable as n is greater than 0.10. This is consistent with the calculations about the surface adsorption of $Bi_2O_2CO_3$ by Chang [34], who suggested that S^{2-} can be adsorbed in the oxygen vacancy of the (001) plane via the chemical bonding and reduce the surface energy. The calculation of the density of states near the Fermi level shows that the doping of S can introduce a new energy level in the energy band and reduce the band gap. The electron state density near the Fermi surface is more diffusive, which favours the migration of electrons and therefore improves the photocatalytic performance.

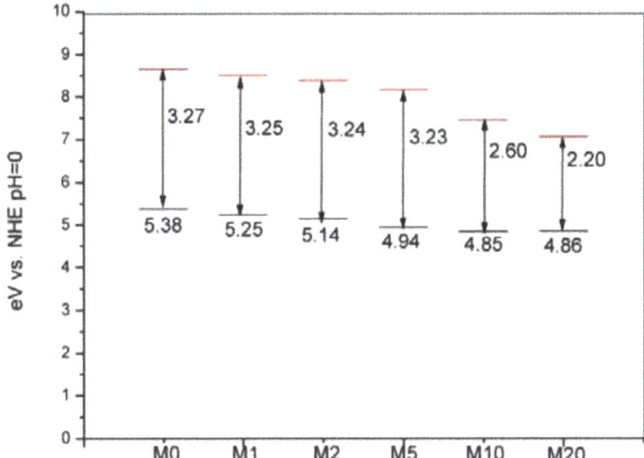

Figure 7. Energy levels of the conduction band minimum (E_{CB}, red) and the valence band maximum (E_{VB}, black) calculated at theoretical pH = 0 (V is voltage; NHE is normal hydrogen electrode potential).

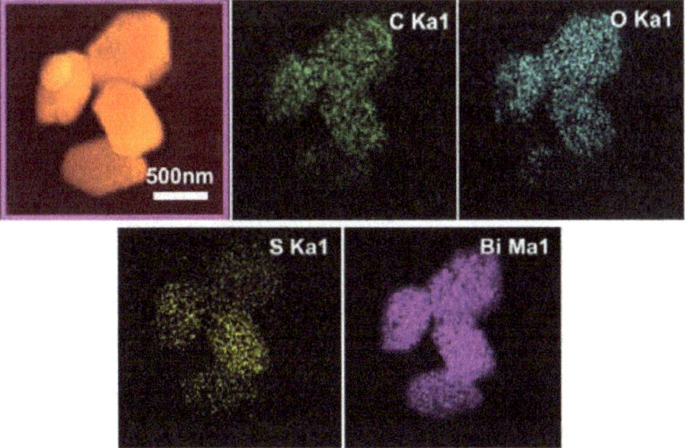

Figure 8. EDS element mapping details of M20 particles.

Table 2. EDS quantitative results of the M0~M20 powders.

	M0	M1	M2	M5	M10	M20
C K	43.78	43.88	43.90	44.11	44.30	44.41
O K	45.53	45.30	45.30	45.27	45.17	45.10
S K	–	0.04	0.06	0.11	0.24	0.29
Bi M	10.69	10.78	10.74	10.51	10.29	10.20
x	0	0.007	0.011	0.021	0.047	0.057

2.2. Ultraviolet-Visible Light Photocatalytic Properties of $Bi_2O_2(CO_3)_{1-x}S_x$

The photocatalytic activity of $Bi_2O_2(CO_3)_{1-x}S_x$ was characterised by photocatalytic degradation of Congo Red. As is shown in Figure 9, the introduction of S could improve the photocatalytic activity of $Bi_2O_2CO_3$ under visible light and UV light. We measured the dye adsorption before switching on the light and normalized the concentrations, which made initial values of c/c_0 equal to 1 for all samples. The operation temperature used was around 0 °C. With pure $Bi_2O_2CO_3$, the Congo Red degrades by 41.6% under the irradiation of visible light for 3h, and by 46.1% under that of UV light, respectively. With the increase of molar ratio of S: $Bi_2O_2CO_3$ from 0.01 to 0.1, the degradation rate increases to 64.2% and 70.1%, respectively. The further increase of n, however, cannot further remarkably increase the degradation rate. At the highest molar ratio of $S(0.2)$, the photocatalytic activity of Congo Red was 65.3% and 71.4%, respectively, which was 1.57 and 1.55 times higher than that of $Bi_2O_2CO_3$, respectively. The photo-degradation behavior of CR by use of $Bi_2O_2(CO_3)_{1-x}S_x$ obeys pseudo-first-order kinetics. This can be fitted by the Langmuir–Hinshewood model of $\ln(C_0/C) = kt + A$, where k is the reaction rate constant, t is the degradation time and the intercept A is the initial value of $\ln(C_0/C)$, which means the dark adsorption of substrates. The k value of M1, M2, M5, M10 and M20 under UV light is 4.3×10^{-3} min^{-1}, 5.5×10^{-3} min^{-1}, 6.1×10^{-3} min^{-1}, 6.8×10^{-3} min^{-1} and 6.9×10^{-3} min^{-1}, respectively. The k value under visible light is 3.0×10^{-3} min^{-1}, 3.5×10^{-3} min^{-1}, 4.7×10^{-3} min^{-1}, 5.5×10^{-3} min^{-1} and 5.9×10^{-3} min^{-1}, respectively. The strong visible light sensitivity indicates higher utilization efficiency of solar light, making $Bi_2O_2(CO_3)_{1-x}S_x$ a superior photocatalyst than the commercial P25 TiO_2, which has been reported to be hardly able to respond to visible light [37,38].

Chang's theoretical calculations [34] suggest that S can be easily captured and adsorbed by oxygen vacancies on the surface of $Bi_2O_2CO_3$ as formed S^{2-} can partially substitute CO_3^{2-} without forming a second phase, introducing a bend built-in electric field. At the same time, their experiments also confirmed that $Bi_2O_2(CO_3)_{1-x}S_x$ had higher conductivity and better carrier transport performance. The photoluminescence (PL) spectra of different S-substituted $Bi_2O_2(CO_3)_{1-x}S_x$ (Figure 10) show that samples M10 and M20 displayed weaker electron holes and recombination, indicating that the introduction of S can effectively suppress the carrier recombination. The stronger light absorption contributed by the smaller band gap means more photo-induced electron hole generation, and those electron holes showed better separation according to PL spectra. It is worthnoting that M20 has astronger light absorption and smaller bandgap than thatof M10, but that they presentnearly the same photocatalytic activity, which may be caused by the smaller specific surface area (Table 1) and slightlyweaker separation (Figure 10) of M20. Thus, all the three factors helped enhance the photocatalytic performance of $Bi_2O_2CO_3$ in our samples under UV and visible light irradiation.

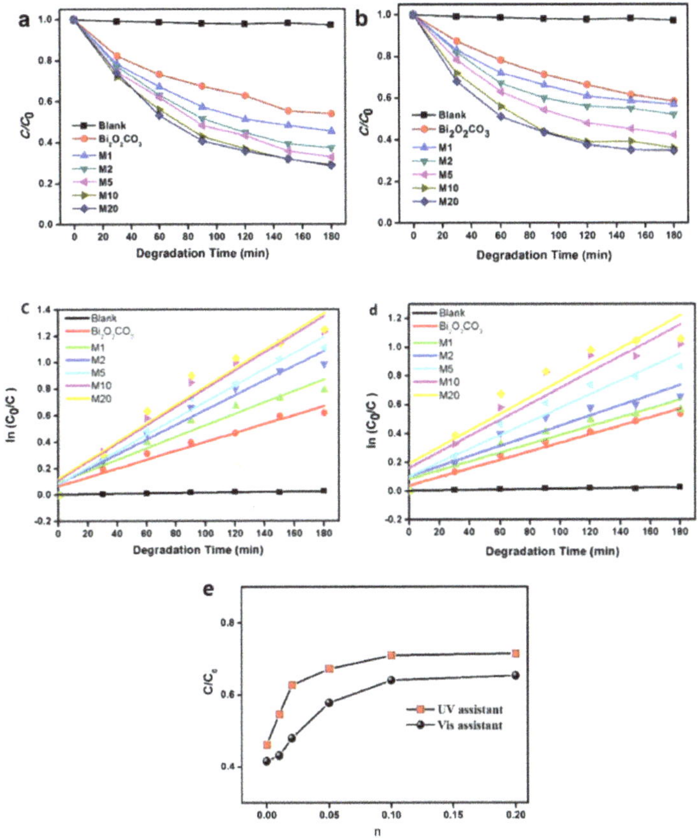

Figure 9. Degradation of Congo Red under the irradiation of (**a**,**c**) UV, and (**b**,**d**) visible light, with $Bi_2O_2(CO_3)_{1-x}S_x$; (**e**) the degradation rate as a function of n (=S:$Bi_2O_2CO_3$).

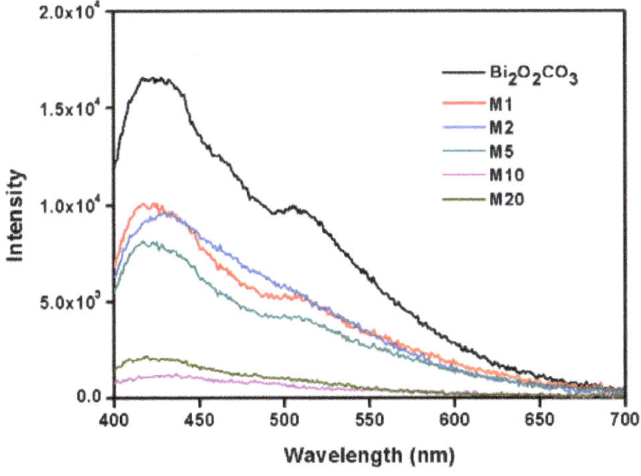

Figure 10. Photoluminescence spectra of $Bi_2O_2(CO_3)_{1-x}S_x$.

3. Materials and Methods

3.1. Preparation of the Precursor Powder $Bi_2O_2CO_3$ via Hydrothermal Method

Three grams of urea ($\geq$99.0%, Beijing Modern Orient Fine Chemistry Co. Ltd., Beijing, China) was dissolved in 60 mL of deionized water in a Teflon hydrothermal tank. 4.65 g Bi_2O_3 powder (99.99%, Aladdin Industrial Corporation, Shanghai, China) was then introduced into the solution. The hydrothermal tank was then tightly closed and kept in an oven at 180 °C for 12 h. After cooling down to room temperature, the precipitate was separated and washed with deionized water and ethanol several times and then dried in the oven at 70 °C.

3.2. Preparation of $Bi_2O_2(CO_3)_{1-x}S_x$ by Chemical Bath Precipitation

Five suspensions of $Bi_2O_2CO_3$ were prepared, each by dispersing 2.04 g of $Bi_2O_2CO_3$ powder in 50 mL of deionized water with the help of ultrasonic stirring for 10 min. A certain amount (S:$Bi_2O_2CO_3$ ratio n, equals to 0.01, 0.02, 0.05, 0.10 and 0.20, respectively) of 0.5 mol/L Na_2S ($\geq$98.0%, Shanghai Tongya Chemical Technology Co. Ltd., Shanghai, China) solution was introduced into the respective suspensions. After 8 h of further magnetic stirring at room temperature, the precipitates were separated and washed several times with deionized water and ethanol and dried at 70 °C. As-treated powders were numbered as M1, M2, M5, M10 and M20, respectively.

3.3. Characterization

Powder X-ray diffraction (XRD) was completed on a diffractometer (D8-Advance, Bruker, Billerica, MA, USA)using monochromatized Cu Kα (λ = 0.15418 nm) radiation with scanning speed of 3°/min. The morphologyof the sampleswerecarried out on a scanning electron microscope (JSM-7001F, JEOL, Tokyo, Japan) operating at a 5 kV and a field emissionelectron microscope (JEM-2100F, JEOL). The surface areas of specimens were tested on a automated gas sorption anslyser (Quantachrome, autosorb iQ2). The X-ray photoelectron spectroscopic (XPS) measurements were performed on a Thermo Fisher ESCALAB 250Xi instrument. A UV-Vis-NIR spectrometer (Lambda 950, PerkinElmer, Waltham, MA, USA) was used to measure UV-Vis diffuse reflectance spectra (DRS). Energy levels of the valence band maximum (E_{VB}) were measured by the ultraviolet photoelectron spectrometer (AC-2, RIKEN KEIKI, Tokyo, Japan).

3.4. Photocatalytic Test

The photocatalytic activity of the prepared $Bi_2O_2(CO_3)_{1-x}S_x$ powder samples was evaluated by photodegrading Congo Red (CR, 100 mg/L) aqueous solution.The reason we chose this concentration is because it is proper to evaluate the change of the color. 0.16 g photocatalyst powder specimen was dispersed into 80 mL CR solution and stirred in the dark for 2 h to reach theadsorption–desorption equilibrium between the photocatalysts and organic dye molecules. Magnetic stirringanda cooling-water bath were held continuously to prevent thermal effect during the degradation process and tokeep the uniformity. A 5W LED with emission wavelength of 365 $\pm$ 5 nm and a 300 W xenon lamp with 420 nm cut-off filters were used as the UV (365~800 nm) and visible light sources (420~800 nm), respectively.The incident light source was placed above the aqueous solution vertically, and the illumination intensity for UV and visible lights at upper surface of the solution were about 78 mW/cm^2 and 132 mW/cm^2. The photocatalytic processes were conducted under constant temperature, using ice water to cool the system. At the end of regular time intervals, 3 mL suspension was collected and centrifuged, and the residual CR concentrationin the supernatant fluid was analyzed by UV-vis spectrophotometer (UV−3100, Hitachi, Tokyo, Japan).

4. Conclusions

AnNa$_2$S chemical bath treatment of Bi$_2$O$_2$CO$_3$ did not generate a second phase. It is shown that the introduction of S can effectively broaden the optical absorption range, although it does not apparently change the crystal structure of Bi$_2$O$_2$CO$_3$. The electrons at the top of the valence band in Bi$_2$O$_2$(CO$_3$)$_{1-x}$S$_x$ can be excited by shorter wavelengths of sunlight, forming photo-generated electron-hole pairs. This may be due to the formation of chemical bonds between the S^{2-} and vacancies on the surface of Bi$_2$O$_2$CO$_3$ crystal, which can affect the surface properties.

Bi$_2$O$_2$(CO$_3$)$_{1-x}$S$_x$ can improve the catalytic performance of visible and UV regions to a certain extent by the introduction of S in Bi$_2$O$_2$CO$_3$ by chemical bath. This is because the introduction of S can effectively suppress the carrier recombination and improve the carrier transport performance. However, S can be introduced only into the surface of Bi$_2$O$_2$CO$_3$ by chemical bath at room temperature, and the improvement of catalytic performance is limited.

Author Contributions: Y.L. and J.D. designed the project; J.D. conducted the most experiments; L.Q. and H.X. performed party of the experiments; J.D. and H.W. together wrote and revised the manuscript with input from all the authors. C.N. and Y.L. took part in discussing and the revision of the manuscript.

Acknowledgments: This work was supported by National Natural Science Foundation of China (no. 51221291, no. 51328203 and 51532003).

Conflicts of Interest: The authors declare no conflict of interest.

References

1. Tong, H.; Ouyang, S.; Bi, Y.; Umezawa, N.; Oshikiri, M.; Ye, J. Nano-photocatalytic materials: Possibilities and challenges. *Adv. Mater.* **2012**, *24*, 229–251. [CrossRef] [PubMed]

2. Wang, P.; Huang, B.B.; Dai, Y.; Whangbo, M.H. Plasmonic photocatalysts: Harvesting visible light with noble metal nanoparticles. *Phys. Chem. Chem.* **2012**, *14*, 9813–9825. [CrossRef] [PubMed]

3. Wang, H.L.; Zhang, L.S.; Chen, Z.G.; Hu, J.Q.; Li, S.J.; Wang, Z.H.; Liu, J.S.; Wang, X.C. Semiconductor heterojunction photocatalysts: Design, construction, and photocatalytic performances. *Chem. Soc. Rev.* **2014**, *43*, 5234–5244. [CrossRef] [PubMed]

4. Moniz, S.J.A.; Shevlin, S.A.; Martin, D.J.; Guo, Z.X.; Tang, J.W. Visible-light driven heterojunction photocatalysts for water splitting—A critical review. *Energy Environ. Sci.* **2015**, *8*, 731–759. [CrossRef]

5. Yu, J.G.; Low, J.X.; Xiao, W.; Zhou, P.; Jaroniec, M. Enhanced photocatalytic CO$_2$-reduction activity of anatase TiO$_2$ by coexposed {001} and {101} facets. *J. Am. Chem. Soc.* **2014**, *136*, 8839–8842. [CrossRef] [PubMed]

6. Chava, R.K.; Do, J.Y.; Kang, M. Hydrothermal growth of two dimensional hierarchical MoS$_2$ nanospheres on one dimensional CdS nanorods for high performance and stable visible photocatalytic H$_2$ evolution. *Appl. Surf. Sci.* **2018**, *433*, 240–248. [CrossRef]

7. Chava, R.K.; Do, J.Y.; Kang, M. Fabrication of CdS-Ag$_3$PO$_4$ heteronanostructures for improved visible photocatalytic hydrogen evolution. *J. Alloys Compd.* **2017**, *727*, 86–93. [CrossRef]

8. Ye, L.Q.; Su, Y.R.; Jin, X.L.; Xie, H.Q.; Zhang, C. Recent advances in BiOX (X = Cl, Br and I) photocatalysts: Synthesis, modification, facet effects and mechanisms. *Environ. Sci. Nano* **2014**, *1*, 90–112. [CrossRef]

9. Xiao, X.; Liu, C.; Hu, R.; Zuo, X.; Nan, J.; Li, L.; Wang, L. Oxygen-rich bismuth oxyhalides: Generalized one-pot synthesis, band structures and visible light photocatalytic properties. *J. Mater. Chem.* **2012**, *22*, 22840–22843. [CrossRef]

10. Kim, T.W.; Choi, K.S. Nanoporous BiVO$_4$ photoanodes with dual-layer oxygen evolution catalysts for solar water splitting. *Science* **2014**, *343*, 990–994. [CrossRef] [PubMed]

11. Kudo, A.; Omori, K.; Kato, H. A novel aqueous process for preparation of crystal form-controlled and highly crystalline BiVO$_4$ powder from layered vanadates at room temperature and its photocatalytic and photophysical properties. *J. Am. Chem. Soc.* **2000**, *31*, 11459–11467. [CrossRef]

12. Zhou, L.; Jin, C.G.; Yu, Y.; Chi, F.L.; Ran, S.L.; Lv, Y.H. Molten salt synthesis of Bi$_2$WO$_6$ powders with enhanced visible-light induced photocatalytic activities. *J. Alloys Compd.* **2016**, *680*, 301–308. [CrossRef]

13. Zhang, J.; Huang, L.H.; Yang, L.X.; Lu, Z.D.; Wang, X.Y.; Xu, G.L.; Zhang, E.P.; Wang, H.B.; Kong, Z.; Xi, J.H.; et al. Controllable synthesis of Bi_2WO_6 (001)/TiO_2 (001) heterostructure with enhanced photocatalytic activity. *J. Alloys Compd.* **2016**, *676*, 37–45. [CrossRef]

14. Kalithasan, N.; Hari, C.B.; Rajesh, J.T. Photocatalytic efficiency of bismuth oxyhalide (Br, Cl and I) nanoplatesfor RhB dye degradation under LED irradiation. *J. Ind. Eng. Chem.* **2016**, *34*, 146–156.

15. Zhang, H.; Yang, Y.; Zhou, Z.; Zhao, Y.; Liu, L. Enhanced photocatalytic properties in BiOBr nanosheets with dominantly exposed (102) facets. *J. Phys. Chem. C* **2014**, *118*, 14662–14669. [CrossRef]

16. Zhang, X.; Liu, Y.; Zhang, G.; Wang, Y.; Zhang, H.; Huang, F. Thermal decomposition of bismuth oxysulfide from photoelectric Bi_2O_2S to Superconducting $Bi_4O_4S_3$. *ACS Appl. Mater. Interfaces* **2015**, *7*, 4442–4448. [CrossRef] [PubMed]

17. Pacquette, A.L.; Hagiwara, H.; Ishihara, T.; Gewirth, A.A. Fabrication of an oxysulfide of bismuth Bi_2O_2S and its photocatalytic activity in a Bi_2O_2S/In_2O_3 composite. *J. Photochem. Photobiol. A Chem.* **2014**, *277*, 27–36. [CrossRef]

18. Tsunoda, Y.; Sugimoto, W.; Sugahara, Y. Intercalation behavior of *n*-alkylamines into a protonated from of a layered perovskite derived from aurivilius phase $Bi_2SrTa_2O_9$. *Chem. Mater.* **2003**, *15*, 632–635. [CrossRef]

19. Huang, Y.; Wang, W.; Zhang, Q.; Cao, J.J.; Huang, R.J.; Ho, W.K.; Lee, S.C. In Situ fabrication of a-Bi_2O_3/(BiO)$_2CO_3$ nanoplateheterojunctions with tunable optical property and photocatalytic activity. *Sci. Rep.* **2016**, *6*, 23435. [CrossRef] [PubMed]

20. Ni, Z.L.; Sun, Y.J.; Zhang, Y.X.; Dong, F. Fabrication, modification and application of (BiO)$_2CO_3$-based photocatalysts: A review. *Appl. Surf. Sci.* **2016**, *365*, 314–335. [CrossRef]

21. Cheng, H.F.; Huang, B.B.; Yang, K.S.; Wang, Z.Y.; Qin, X.Y.; Zhang, X.Y.; Dai, Y. Facile template-free synthesis of $Bi_2O_2CO_3$ hierarchical microflowers and their associated photocatalytic activity. *J. Phys. Chem. C* **2010**, *11*, 2167–2173. [CrossRef] [PubMed]

22. Dong, F.; Xiong, T.; Sun, Y.J.; Huang, H.W.; Wu, Z.B. Synergistic integration of thermocatalysis and photocatalysis on black defective (BiO)$_2CO_3$ microspheres. *J. Mater. Chem. A* **2015**, *3*, 18466–18474. [CrossRef]

23. Huang, H.W.; Wang, J.J.; Dong, F.; Guo, Y.X.; Tian, N.; Zhang, Y.H.; Zhang, T.R. Highly Efficient $Bi_2O_2CO_3$ Single-Crystal Lamellas with Dominantly Exposed {001} Facets. *Cryst. Growth Des.* **2015**, *15*, 534–537. [CrossRef]

24. Huang, H.W.; Tian, N.; Jin, S.F.; Zhang, Y.H.; Wang, S.B. Syntheses, characterization and nonlinear optical properties of a bismuth subcarbonate $Bi_2O_2CO_3$. *Solid State Sci.* **2014**, *30*, 1–5. [CrossRef]

25. Madhusudan, P.; Ran, J.R.; Zhang, J.; Yu, J.G.; Liu, G. Novel urea assisted hydrothermal synthesis of hierarchical $BiVO_4$/$Bi_2O_2CO_3$nanocomposites with enhanced visible-light photocatalytic activity. *Appl. Catal. B* **2011**, *110*, 286–295. [CrossRef]

26. Liang, N.; Zai, J.T.; Xu, M.; Zhu, Q.; Wei, X.; Qian, X.F. Novel Bi_2S_3/$Bi_2O_2CO_3$ heterojunction photocatalysts with enhanced visible light responsive activity and wastewater treatment. *J. Mater. Chem. A* **2014**, *2*, 4208–4216. [CrossRef]

27. Huang, Y.; Fan, W.; Long, B.; Li, H.; Zhao, F.; Liu, Z.; Tong, Y.; Ji, H. Visible light Bi_2S_3/Bi_2O_3/$Bi_2O_2CO_3$ photocatalyst for effective degradation of organic pollutions. *Appl. Catal. B Environ.* **2016**, *185*, 68–76. [CrossRef]

28. Dong, F.; Li, Q.Y.; Sun, Y.J.; Ho, W.K. Noble metal-like behavior of plasmonic Bi particles as a cocatalyst deposited on (BiO)$_2CO_3$ microspheres for efficient visible light photocatalysis. *ACS Catal.* **2014**, *4*, 4341–4350. [CrossRef]

29. Xiong, T.; Huang, H.W.; Sun, Y.J.; Dong, F. In-Situ synthesis of a C-doped (BiO)$_2CO_3$ hierarchical self-assembly effectively promoting visible light photocatalysis. *J. Mater. Chem. A* **2015**, *3*, 6118–6127. [CrossRef]

30. Huang, H.W.; Xiao, K.; Yu, S.X.; Dong, F.; Zhang, T.R.; Zhang, Y.H. Iodide surface decoration: An facile and efficacious approach to modulating the band energy level of semiconductors for high performance visible-light Photocatalysis. *Chem. Commun.* **2016**, *52*, 354–357. [CrossRef] [PubMed]

31. Dong, F.; Li, Q.; Ho, W.; Wu, Z.B. The mechanism of enhanced visible light photocatalysis with micro-structurally optimized and graphene oxide coupled (BiO)$_2CO_3$. *Chin. Sci. Bull.* **2015**, *60*, 1915–1923.

32. Li, Q.; Liu, H.; Dong, F.; Fu, M. Hydrothermal formation of N-doped (BiO)$_2CO_3$ honeycomb-like microspheres photocatalysts with bismuth citrate and Dicyandiamide as precursors. *J. Colloid Interface Sci.* **2013**, *408*, 33–42. [CrossRef] [PubMed]

33. Liu, Y.; Wang, Z.; Huang, B.; Yang, K.; Zhang, X.; Qin, X.; Dai, Y. Preparation, electronic structure, and photocatalytic properties of $Bi_2O_2CO_3$ nanosheet. *Appl. Surf. Sci.* **2010**, *257*, 172–175. [CrossRef]

34. Chang, C.; Teng, F.; Liu, Z. Fully Understanding the Photochemical Properties of $Bi_2O_2(CO_3)_{1-x}S_x$ Nanosheets. *Langmuir* **2016**, *32*, 3811–3819. [CrossRef] [PubMed]

35. Huang, H.; Li, X.; Wang, J.; Dong, F.; Chu, P.K.; Zhang, T.; Zhang, Y. Anionic Group Self-Doping as a Promising Strategy: Band-Gap Engineering and Multi-Functional Applications of High-Performance CO_3^{2-}-Doped $Bi_2O_2CO_3$. *ACS Catal.* **2015**, *5*, 4094–4103. [CrossRef]

36. Zhao, Z.; Zhou, Y.; Wang, F.; Zhang, K.; Yu, S.; Cao, K. Polyaniline-Decorated {001} Facets of $Bi_2O_2CO_3$ Nanosheets: In Situ Oxygen Vacancy Formation and Enhanced Visible Light Photocatalytic Activity. *ACS Appl. Mater. Interfaces* **2015**, *7*, 730–737. [CrossRef] [PubMed]

37. Pawar, R.C.; Khare, V.; Lee, C.S. Hybrid photocatalysts using graphitic carbon nitride/cadmium sulfide/reduced graphene oxide (g-C_3N_4/CdS/RGO) for superior photodegradation of organic pollutants under UV and visible light. *Dalton Trans.* **2014**, *43*, 12514–12527. [CrossRef] [PubMed]

38. Liu, G.; Chen, Z.; Dong, C.; Zhao, Y.; Li, F.; Lu, G.Q.; Cheng, H.-M. Visible Light Photocatalyst: Iodine-Doped Mesoporous Titania with a Bicrystalline Framework. *J. Phys. Chem. B* **2006**, *110*, 20823–20828. [CrossRef] [PubMed]

 © 2018 by the authors. Licensee MDPI, Basel, Switzerland. This article is an open access article distributed under the terms and conditions of the Creative Commons Attribution (CC BY) license (http://creativecommons.org/licenses/by/4.0/).

Article

New Insights on the Photodegradation of Caffeine in the Presence of Bio-Based Substances-Magnetic Iron Oxide Hybrid Nanomaterials

Davide Palma [1,2] [ID], Alessandra Bianco Prevot [1,*][ID], Marcello Brigante [2], Debora Fabbri [1] [ID], Giuliana Magnacca [1,3] [ID], Claire Richard [2], Gilles Mailhot [2] and Roberto Nisticò [4,*][ID]

[1]   Department of Chemistry, University of Torino, via P. Giuria 7, 10125 Torino, Italy; davide.palma@etu.uca.fr (D.P.); debora.fabbri@unito.it (D.F.); giuliana.magnacca@unito.it (G.M.)
[2]   CNRS, SIGMA Clermont, Institut de Chimie de Clermont-Ferrand, Université Clermont Auvergne, F-63000 Clermont-Ferrand, France; marcello.brigante@uca.fr (M.B.); claire.richard@uca.fr (C.R.); gilles.mailhot@uca.fr (G.M.)
[3]   NIS (Nanostructured Interfaces and Surfaces) Centre, Via P. Giuria 7, 10125 Torino, Italy
[4]   Polytechnic of Torino, Department of Applied Science and Technology DISAT, C.so Duca Degli Abruzzi 24, 10129 Torino, Italy
*   Correspondence: alessandra.biancoprevot@unito.it (A.B.P.); roberto.nistico@polito.it (R.N.); Tel.: +39-011-670-5292 (A.B.P.); +39-011-090-4762 (R.N.)

Received: 25 May 2018; Accepted: 22 June 2018; Published: 26 June 2018

Abstract: The exploitation of organic waste as a source of bio-based substances to be used in environmental applications is gaining increasing interest. In the present research, compost-derived bio-based substances (BBS-Cs) were used to prepare hybrid magnetic nanoparticles (HMNPs) to be tested as an auxiliary in advanced oxidation processes. Hybrid magnetic nanoparticles can be indeed recovered at the end of the treatment and re-used in further water purification cycles. The research aimed to give new insights on the photodegradation of caffeine, chosen as marker of anthropogenic pollution in natural waters, and representative of the contaminants of emerging concern (CECs). Hybrid magnetic nanoparticles were synthetized starting from Fe(II) and Fe(III) salts and BBS-C aqueous solution, in alkali medium, via co-precipitation. Hybrid magnetic nanoparticles were characterized via X-ray diffraction (XRD), thermo-gravimetric analysis (TGA) and Fourier transform infrared (FTIR) spectroscopy. The effect of pH, added hydrogen peroxide, and dissolved oxygen on caffeine photodegradation in the presence of HMNPs was assessed. The results allow for the hypothesis that caffeine abatement can be obtained in the presence of HMNPs and hydrogen peroxide through a heterogeneous photo-Fenton mechanism. The role of hydroxyl radicals in the process was assessed examining the effect of a selective hydroxyl radical scavenger on the caffeine degradation kinetic.

Keywords: advanced oxidation processes; bio-based substances; hybrid nanomaterials; magnetic materials; photo Fenton; caffeine

---

1. Introduction

Polluted water treatment as well as urban bio-waste (UBW) management represent two challenging key issues that have to be faced worldwide. On one side, access to water has been recognized as a fundamental human right by the United Nations General Assembly [1], and "clean water and sanitation" is one of the United Nations Sustainable Development Goals [2,3]. Water scarcity, poor water quality, and inadequate sanitation negatively impact millions of people across the world every year. Moreover, the way water is used, which can mainly be described as "linear" (that is extraction upstream, disinfection treatment processes, use and then application of more expensive treatment processes before discharging it downstream) has numerous malfunctions that threaten the health of people and the environment. Therefore, there is a rising demand for exploring integrated water use and water treatment processes seeking for a transition to a more circular water management [4]. A peculiar aspect related to water quality is represented by the so-called contaminants of emerging concern (CECs)—xenobiotics detected at very low level in natural water bodies, since their recalcitrant behavior is to be abated in (waste)water treatment. Contaminants of emerging concern span a variety of chemicals comprising pharmaceuticals and personal care products (PPCPs), endocrine-disrupting compounds (EDCs), flame retardants (FRs), pesticides, and artificial sweeteners (ASWs) and their metabolites (see References [5–7] and references therein). Based on their harmfulness to human and environment health, there is an increasing commitment to find efficient processes for their abatement [8–14]. A CEC abatement approach that has been proposed by several research groups is represented by the so-called advanced oxidation processes (AOPs) that exploit the generation of highly reactive species (mainly HO· radicals) to induce the oxidative degradation of the organic pollutants, until their complete mineralization is attained. Among AOPs, great and increasing interest has been devoted to both Fenton and photo-Fenton processes, with a particular emphasis on the possibility to implement the process at a mild circumneutral pH, which is more compatible with natural water bodies [15–20].

On the other side, the increasing environmental concern of our society with respect to UBW fate has resulted in the need for developing sustainable processes able to decrease waste production or, alternatively, to valorize them for other uses. This latter approach fits with the biorefinery concept that is integrating different biomass conversion approaches to obtain fuels, power, heat, and value-added chemicals. Materials such as compost, anaerobic digestate, and organic residues from mechanical-biological treatment of UBW can be studied as biomass that could be used to feed a biorefinery [21]. It was previously demonstrated that with this approach, it is possible to extract from UBW, anaerobically and/or aerobically treated, bio-based substances (hereinafter BBS) featuring similarly to soil humic substances under the structural and physicochemical point of view [22,23]. Bio-based substances have been tested in several technological and environmental applications, including wastewater treatment [24–26].

Coupling UBW development and water treatment can therefore be considered an interesting way to "close the loop" in the transition from a linear to a circular economy [27].

We previously reported promising performances of BBS derived from compost as an auxiliary in the photodegradation of a group of CECs in the presence of iron and hydrogen peroxide (i.e., in a photo-Fenton process) [28,29]. In the presence of added BBS, it was possible to operate at milder pH conditions, compared to the ones suggested for photo-Fenton pollutant degradation [30].

A further step is represented by moving to a heterogeneous process, where BBS can be used as coating of magnetic nanoparticles made of magnetite/maghemite, containing iron useful for both Fenton and photo-Fenton reactions and suitable to be recovered and reused after water treatment [16,31]. Actually, magnetite/maghemite particles have been proposed for Fenton and photo-Fenton pollutant degradation with promising results (see Reference [32] and references therein). Nevertheless, these oxides undergo oxidation in natural atmosphere, yielding non-magnetic hematite, and the addition of organic coatings during the magnetite synthesis could also act as stabilizing barrier against iron (II) oxidation [13,33]. The addition of BBS was demonstrated to stabilize magnetite/maghemite nanoparticles [34], and in the presence of this kind of materials, encouraging preliminary results have been obtained in the photodegradation of caffeine [35,36]. Indeed, the hybrid BBS-coated magnetic nanoparticles (HMNPs) enhanced the caffeine photodegradation in the presence of added hydrogen peroxide.

In the present paper, more insights are given on the photodegradation of caffeine, a highly-used stimulant contained in drinking products (e.g., coffee, tea, caffeinated beverages) as well as in many pharmaceutical and personal care products. Its presence in the environment has been extensively studied; due to its environmental ubiquity caffeine has become a commonly used anthropogenic marker for water pollution, and it has been included among the contaminants of emerging concern (CECs). Even if caffeine has been traditionally accepted as posing a low risk to aquatic environments, it might grant attention; its mixture of toxicity effects, together with the ability of caffeine to bio-accumulate in the tissues of some aquatic species could represent a potential environmental risk. Thus, several studies have been devoted to assess the environmental hazards posed by caffeine [37].

Compost-derived bio-based substances (BBS-Cs) have been previously obtained from commercially composted urban biowastes [38] following a standard procedure for the humic acid isolation from soils; these composted biowastes are derived from urban public park trimmings and home gardening residues aged for more than 180 days [22]. After isolation, BBS-Cs have been characterized for their physico-chemical features [22]. Compost-derived bio-based substances have been chosen to prepare hybrid magnetic nanoparticles to be tested in the caffeine photodegradation process in order to increase the value of a material that is a product of a "virtuous" waste management approach, but, even if commercially available, does not have a market value suitable for compensating the cost of its production (including the initial separate waste collection). The effect of several experimental variables has been considered and assessed, and the reusability of HMNPs has been checked.

2. Materials and Methods

2.1. Materials

2.1.1. BBS-C Isolation from the Green Compost

Compost-derived bio-based substances were obtained from commercially-composted urban biowastes (*ACEA Pinerolese Industriale S.p.A.*, Pinerolo, Italy) [38]. The BBS-C isolation process was based on a standard procedure for the humic acid isolation from soils, and has been reported in Reference [22]. As shown in Scheme 1, it consists of a three-step process. In the first step, BBS-Cs were extracted from the green compost by alkaline washing with 0.5 M NaOH solution (1:10 solid–solution ratio, Sigma-Aldrich, Saint Louis, MO, USA) overnight in the dark. Then, the BBS-C solution was separated from the solid residues by several centrifugation and filtration steps (diameter: 2.7 μm). In the second step, BBS-Cs were precipitated via flocculation by varying the pH into acid (HCl conc. addition, pH < 1) and leaving the suspension overnight in the dark. Then the solid BBS-Cs were separated from the liquid phase by centrifugation (10,000 rpm, 20 min). In the last step, BBS-Cs were purified from NaCl residues (derived from the previous three steps) by cyclic washing with deionized water and subsequent dialysis in tubular membranes (cut-off 1 kDa) for 3–4 days rinsing with fresh-deionized water until the conductivity value decreases to 10 μS·cm^{-1}. Finally, the suspension was freeze-dried, and solid BBS-C stocked.

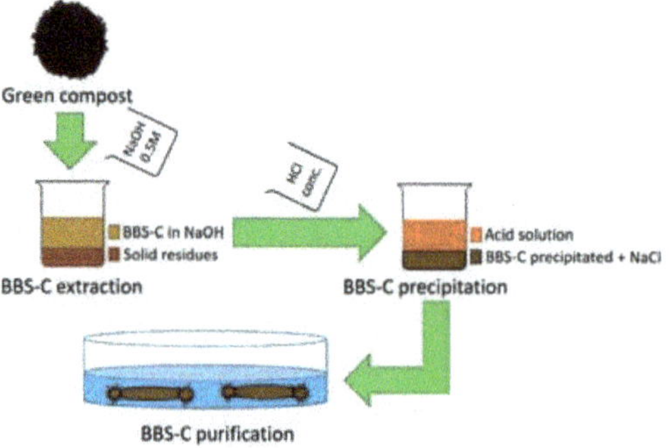

Scheme 1. Schematic representation of compost-derived bio-based substance (BBS-C) isolation from compost.

2.1.2. HMNPs Synthesis

The iron oxide magnetic nanoparticles were produced following a modified procedure based on a consolidated co-precipitation process reported in our previous studies [16,26,35,36,39]. In detail, 6.16 g $FeCl_3 \cdot 6H_2O$ (purity = 98%, CAS 10025-77-1, Sigma-Aldrich, Saint Louis, MO, USA) and 4.2 g $FeSO_4 \cdot 7H_2O$ (purity > 99.0%, CAS 7782-63-0, Sigma-Aldrich) were dissolved into 100 mL of deionized water. The solution was heated up to 90 °C under mechanical stirring. Afterwards, two solutions were added simultaneously: (i) 10 mL of an ammonia solution (30%, CAS 7664-41-7, Sigma-Aldrich), and (ii) 50 mL of a BBS-C solution containing the desired amount of BBS-C. Three different BBS-C solutions were investigated: 0.1, 0.2, and 1.0 wt.%. With respect to our previous work [36], smaller amounts of BBS-C were used to functionalize the magnetic material in order to evidence the effects of the experimental parameters examined and obtain information on material stability and reaction mechanisms. After 30 min at isothermal conditions (90 °C), nanoparticles obtained were magnetically separated using a commercial neodymium magnet, washing 5–6 times with fresh deionized water in order to remove the by-products of the reaction (i.e., ammonium salts). Finally, the BBS-Cs were deposited onto a glass Petri dish and dried at 80 °C overnight. Dried particles were crushed in a mortar. Samples thus obtained were named MH0.1, MH0.2, and MH1.0, depending on the BBS-C content in the solution used during the synthesis. Additionally, a reference material of pure magnetic iron oxide (coded M0) was synthesized in the absence of BBS-C solution.

2.1.3. Other Reagents

Sodium hydroxide (NaOH, purity $\geq$ 98.0%, CAS 1310-73-2, Sigma-Aldrich), hydrochloric acid (HCl, conc. 37 wt.%, CAS 7647-01-0, Sigma-Aldrich), anhydrous potassium chloride (KCl, purity $\geq$ 99.0%, CAS 7447-40-7, Fluka), caffeine (CAS 58-08-2, Sigma-Aldrich), H_2O_2 (v/v 30%, CAS 7722-84-1, Sigma-Aldrich), *orto*-phenanthroline (purity 99.0%, CAS 66-71-7, Sigma-Aldrich), methanol (HPLC grade, $\geq$99.9%, CAS 67-56-1, Sigma-Aldrich), and 2-propanol (purity 95%, CAS 67-63-0, Sigma-Aldrich) were used throughout the work. All aqueous solutions were prepared using ultrapure water Millipore Milli-Q™ (Merk, Darmstadt, Germany). All chemicals were used without further purification.

2.2. Methods

2.2.1. HMNPs Characterization

Thermo-gravimetric analyses (TGA) were performed on a TGA Q600 STD (TA Instruments, New Castle, DE, USA), working under either inert (nitrogen) or oxidant (air) atmospheres. The weight losses and the degradation profiles were evaluated by heating ca. 5–10 mg of each sample in an open alumina pan, applying a heating ramp from 30 to 800 °C (rate 10 °C·min^{-1}).

X-ray diffraction (XRD) patterns were collected on the powdery samples using a PW3040/60 X'Pert PRO MPD diffractometer (PANalytical, Malvern, UK), working with a Cu anode source, at 45 kV and 40 mA in a Bragg–Brentano geometry with flat configuration. The Scherrer equation applied to the (311) iron oxide phase reflection was used to estimate the average size of the crystalline domains.

$$\tau = \frac{K\lambda}{\beta \cos \theta} \tag{1}$$

where, τ is the average size of the crystalline domains (in nm), K is the shape factor (a constant value, for quasi-spherical domains is 0.9), λ is the X-ray wavelength (for Cu source is 0.154 nm), β is the line broadening at half of the maximum intensity after subtracting the line broadening (radians), and θ is the Bragg angle (radians).

Fourier transform infrared (FTIR) spectroscopy was carried out by using a Vector 22 spectrophotometer (Bruker, Billerica, MA, USA) in transmission mode, working with 128 scans and 4 cm^{-1} of resolution in the 4000–400 cm^{-1} range. The instrument has a Globar source and a DTGS detector. Spectra were analyzed by dispersing the samples in KBr (1:20 wt. ratio) and pressing them, forming a homogeneous pellet.

Orto-phenanthroline colorimetric tests were performed on the aqueous phase of a dispersion of MH1.0 HMNPs to evaluate the iron released by the material. Tests were performed on three different aqueous phases of MH1.0 dispersion (concentration 200 mg L^{-1}) at three different pH values (3.0, 5.0, and 7.0). The pH was adjusted by adding either HCl or NaOH diluted solutions. Suspensions were mechanically shacked overnight in the dark. Then HMNPs were separated by applying an external magnetic field by means of a neodymium magnet. The three supernatants were filtered at 0.45 μm and analyzed in a double-beam T90+ UV–vis spectrometer (PG Instruments Ltd., Lutterworth, UK), in a quartz cuvette, slow speed mode at 1 nm resolution in the 200–800 nm range. The colorimetric test was performed as follows: 4 mL of the supernatant were added with 1 mL of 0.1% *w/v* o-phenanthroline aqueous solution and 1 mL of phosphate buffer (pH = 4). In the presence of Fe(II) ions, the orange-red ferrous-tris-*orto*-phenanthroline complex was formed (λ_{max} = 510 nm). The quantification was realized by means of an external calibration (i.e., Fe(II) standards concentration 0.1, 0.5, 1.0, and 5.0 mg·L^{-1}). Additionally, in order to evaluate the total iron amount (i.e., Fe(II) + Fe(III)), a spatula tip of ascorbic acid was added to the supernatant prior to the colorimetric test; this way also Fe(III) was reduced to Fe(II) [40], allowing the determination of the total iron concentration.

The stability of the organic matter at the HMNPs' surface was evaluated on MH0.2 sample (200 mg·L^{-1}) put in contact for 2 h with either deionized water or a 1.5 × 10^{-3} M H_2O_2 solution in the dark or under irradiation (340 nm). The presence of BBS-C fragments released from the HMNPs has been evaluated by UV–Vis spectroscopy considering the absorbance at λ = 254 and 365 nm, values reported in the literature as characteristic for the humic acids [41], and at which the original BBS-C aqueous solution absorbs.

2.2.2. Caffeine Irradiation Test and Analysis

All the irradiation experiments were performed in a cylindrical reactor equipped with 6 Philips TL D 15W/05 tubular lamps having emission spectrum from 300 to 475 nm with a maximum centered at 365 nm. Two-hundred mL of solution were placed in a cylindrical quartz reactor with 4.5 cm internal diameter and maintained under mechanical stirring during the irradiation. Caffeine concentration was fixed at 5 mg·L^{-1} while HMNPs and H_2O_2, when used, were added at a concentration of 200 mg·L^{-1} and 1.5 × 10^{-3} M, respectively.

Caffeine concentration during the irradiation experiments was monitored using a Waters ACQUITY UPLC system (Waters S.p.A., Sesto San Giovanni (MI), Italy) with a Nucleodur C_{18} column (100 mm × 2 mm × 1.8 μm), runtime was 6 min long, and eluents were acidic MilliQ water (0.1% phosphoric acid) and acetonitrile with a flow rate of 0.2 mL·min^{-1}. A gradient raising the percentage of ACN from 40% to 70% during the run was applied. The caffeine quantification wavelength was 260 nm, and retention time was 3.6 min. Before analysis, every sample containing HMNPs were magnet-cleaned (in order to avoid the presence of the magnetic nanoparticles) and filtered through PTFE membranes with 0.45 μm cut-off. In the experiments where H_2O_2 was used, all samples were spiked with methanol with a volume ratio sample of MeOH equal to 1:1 in order to inactivate a Fenton reaction in dark conditions; the concentration values of caffeine were corrected according to the dilution.

3. Results

3.1. HMNPs Main Physico-Chemical Features

The thermal stability and the organic matter content for all HMNPs were evaluated by TGA analyses performed under either reducing (nitrogen) or oxidant atmospheres (see Figure 1A,B, respectively). The thermal profiles obtained under a nitrogen atmosphere (Figure 1A) shows a first weak weight loss at ca. 100 °C due to the water molecules being physically sorbed at the surface, followed by a complex one in the 150–500 °C range corresponding to the pyrolysis of BBS-C, organized in a first contribution centered at 310 °C (only MH1.0) attributable to the carbohydrate residues, and the main one centered at ca. 430 °C due to the pyrolysis of BBS-C humic-like fraction, before forming a carbonaceous residue (BBS-C residue at 800 °C is ca. 40 wt.%) [35]. Once the pyrolysis temperature became higher than 700 °C, a weak weight variation was also registered for both the reference M0 and the HMNPs due to the iron oxide reduction to wustite (FeO) and elemental iron (Fe), in accordance with our previous studies [26,31]. When HMNPs were heated under an oxidant atmosphere (Figure 1B), three weight losses were registered: the first one (at ca. 100 °C), again due to the moisture content in the samples, whereas the main relevant one was comprised in the 250–400 °C range, and it was due to the oxidation of BBS-C to CO_2 and other volatile products. This is since the atmosphere of oxidant BBS-C when thermally treated is completely mineralized leaving almost no residue at 600 °C. At ca. 580 °C (i.e., when the organic matter has been completely oxidized), a phase transformation from magnetite/maghemite to hematite started, thus leaving a reddish non-magnetic residue. Weight losses (calculated in the 150–500 °C range) associated with the organic matter content in the HMNPs were 2 wt.% for MH0.1, 7wt.% for MH0.2, and 15 wt.% for MH1.0, thus suggesting that higher BBS-C concentration during synthesis caused higher BBS-C loading in the nanomaterials.

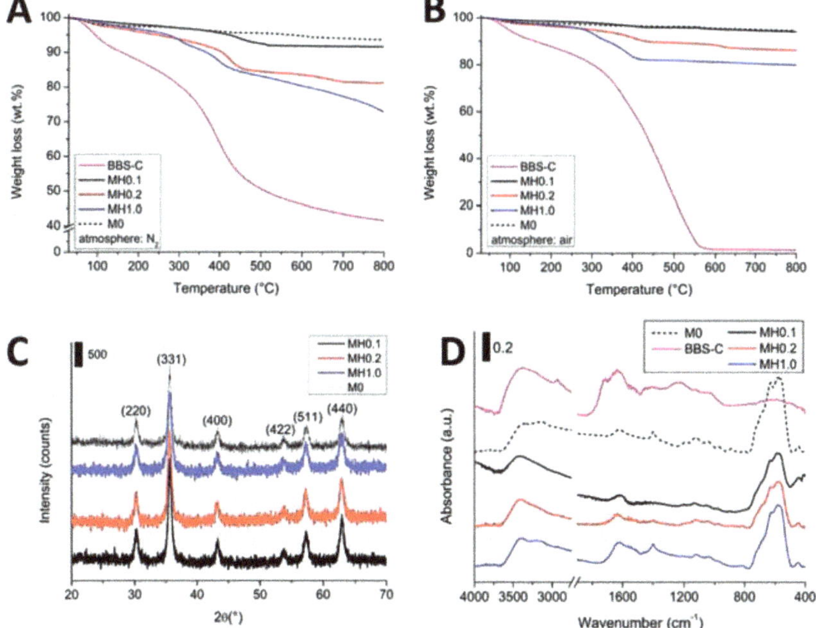

Figure 1. Physicochemical characterization of hybrid magnetic nanoparticles (HMNPs) and reference materials (BBS-C and M0): (**A**) thermo-gravimetric analysis (TGA) under nitrogen atmosphere, (**B**) TGA under air atmosphere; (**C**) X-ray diffraction (XRD) pattern of HMNPs (main reflections are labeled), and (**D**) absorbance FTIR spectra in transmission mode. Legend: M0 (black dotted line), BBS-C (magenta solid line), MH0.1 (black solid line), MH0.2 (red solid line), and MH1.0 (blue solid line).

The X-ray diffraction (Figure 1C) patterns for all HMNPs confirmed the crystalline reflections at 2θ = 30.1° (220), 35.4° (311), 43.0° (400), 53.9° (422), 57.2° (511), and 62.6° (440), consistent with the signals of the reference magnetite/maghemite M0 (card numbers 00-019-0629 and 00-039-1346, ICCD Database) [13,35]. The two phases (namely, magnetite and maghemite) cannot be distinguished by means of XRD, since both phases have the same identical XRD pattern. The only difference is that maghemite is the (topotactic) oxidized form of magnetite, and it is a common procedure to consider the simultaneous presence of both phases [13,35]. Applying the Scherrer equation on the main reflection peak at (311), it is possible to estimate the average size of the crystalline domains, being in the 16–18 nm range for all HMNPs. No interesting peaks were registered for BBS-C since it was amorphous. The exact composition of the BBS-C is very difficult to clarify due to the high complexity of humic substances. As reported in our previous work [22], these BBS-Cs are supramolecular aggregates made by macromolecules containing aromatic and aliphatic chains with lots of functionalities (mostly, O-containing groups). On the basis of the FTIR analysis reported in Figure 1D, the main signals of BBS-C are the OH stretching mode at 3400 cm^{-1}, the CH stretching vibrations in the 3000–2800 cm^{-1}, the C=O stretching mode of the carbonyl functionalities at ca. 1740–1700 cm^{-1}, the signal at ca. 630–550 cm^{-1} due to the Fe-O vibration mode of both the octahedral and tetrahedral sites forming the magnetite/maghemite iron oxide, and several broad signals in the 1650–1560 cm^{-1} range due to C=C vibrations [36,41]. The attribution of the bands at 1400 cm^{-1} deserves more attention. Indeed, significant differences in the spectra bands related to carboxylic and carboxylate groups can be observed between the free BBS-C, M0, and hybrid MH samples. In particular, the pointed peak at 1400 cm^{-1} observed for the MH1.0 sample is consistent with a carboxylate–iron stretching, as previously reported [35,36]. This characteristic signal decreases in the MH0.2 spectrum, and is not detectable in the MH0.1 spectrum. Nevertheless, we have to take into account that the organic matter amount decreases going from MH1.0 to MH0.1, and the absence of the signal ascribed to the BBS-O-Fe could be just a matter of detection limit. An alternative hypothesis could be that the signal at 1400 cm^{-1} was due to the CO bond stretching mode of the carbonate groups formed from the atmospheric CO_2; nevertheless, we discarded this latter hypothesis since the signal intensity should not vary among the different HMNPs.

Reference M0 shows the main signal at ca. 630–550 cm^{-1} (see comment above). On the basis of the FTIR analysis, all HMNPs present the signals due to iron oxide (Fe-O bands), and some peaks attributable to BBS-C-derived organic matter at the surface forming the HMNPs [35].

3.2. HMNPs Stability

In order to evaluate the stability of the HMNPs systems, spectroscopic analyses were performed to quantify the iron released in water (at different pH values), and the release of BBS-C at the experimental conditions. Concerning the release of both Fe(II) and Fe(III) ions in water, a suspension of MH1.0 (200 mg·L^{-1}) was taken as reference for all the HMNPs. A colorimetric test was performed (see Section 2.2.1) after 12 h of HMNP dispersion in ultrapure water. Three different pH conditions were investigated, namely, pH = 3.0, 5.0, and 7.0. Results obtained evidenced that at strong acid pH (pH = 3.0) there was a significant release of Fe(II) from the iron oxide (Table 1). This effect is in agreement with the literature since magnetite/maghemite are sensible to strong acid [42]. On the contrary, at pH 5.0 and 7.0, a significantly higher stability of the materials, in terms of iron release, was observed.

Table 1. Fe(II) and Fe(III) concentrations in water at different pH values from MH1.0 (200 mg·L^{-1}) suspension after 12 h.

pH	Fe(II) (mg·L^{-1})	Fe(III) (mg·L^{-1})	Fe Tot. (mg·L^{-1})
3.0	0.837	0.153	0.993
5.0	0.036	0.014	0.049
7.0	0.137	0.051	0.189

The eventual release of organic matter from HMNPs in solution was evaluated by measuring the UV–Vis spectra of the aqueous phase of MH0.2 (200 mg·L^{-1}) suspensions after 2 h (either in the dark or upon light irradiation) with ultrapure water or with a 1.5×10^{-3} M H_2O_2 solution. As reported in Figure 2, neither the aqueous phase from the irradiated MH0.2 suspension, nor the one from the suspension left in the dark, showed any significant signals in the 190–700 nm wavelength range. Conversely, for the suspensions added by H_2O_2, the corresponding aqueous phases presented an absorption band at $\lambda < 300$ nm assignable to BBS-C (and/or its fragments) released from the HMNPs upon chemical oxidation (in the dark) and/or photochemical oxidation (under irradiation). The very slow but not negligible photodegradation of BBS was reported in a previous study, on the BBS photostability in a homogeneous system [43].

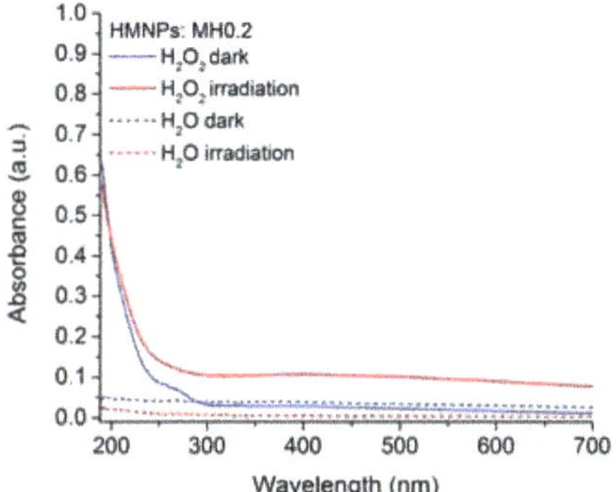

Figure 2. HMNP stability in solution: UV–Vis spectra of either water (dotted line) or 1.5×10^{-3} M H_2O_2 (solid line) solutions after being in contact with MH0.2 for 2 h either in the dark (blue) or under irradiation (red).

3.3. Preliminary Test on Caffeine (Photo)stability

The (photo)stability of caffeine has been preliminarily assessed in homogeneous systems by irradiating an aqueous solution at different experimental conditions: (i) without any additional substance, (ii) in the presence of 1.5×10^{-3} M added H_2O_2, and (iii) in the presence of 6 mg·L^{-1} of BBS-C (theoretical concentration corresponding to BBS-C immobilized on the HMNPs, estimated by TGA analysis). Tests (ii) and (iii) were performed also in the dark.

From data reported in Figure 3, it can be observed that caffeine was not degraded in pure water, and also upon addition of BBS-C the caffeine abatement was negligible. As for this latter experiment, it must be noted that a very low amount of BBS-C was added, compared to the values reported in the literature in other applications of BBS to organic pollutant degradation [24,35].

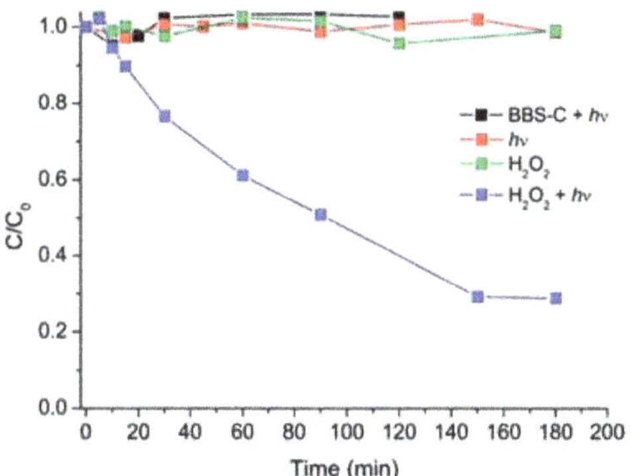

Figure 3. Caffeine concentration ($C_0 = 5$ mg·L^{-1}) profile in the presence of H_2O_2 (1.5×10^{-3} M) in dark condition (green squares); under irradiation (red squares); in the presence of BBS-C (6 mg·L^{-1}), and under irradiation (black squares); in the presence of H_2O_2 (1.5×10^{-3} M), and under irradiation (blue squares).

In the presence of hydrogen peroxide 1.5×10^{-3} M, the caffeine concentration decreased by about 75% after 120 min under irradiation (Figure 3). Despite no degradation in the dark, the fast degradation under irradiation observed in the presence of hydrogen peroxide is attributed to the generation of hydroxyl radicals through the hydrogen peroxide photolysis.

3.4. Heterogeneous Caffeine (Photo)degradation in the Presence of HMNPs

Several preliminary experiments (data not shown) were performed on caffeine aqueous solution using the three synthesized HMNPs: (i) caffeine adsorption on the HMNPs, in the dark and (ii) caffeine degradation under irradiation. All the experiments were run for 180 min and the three HMNPs did not show significantly different behaviors; in both cases (i) and (ii), the caffeine disappearance was negligible. Different results were obtained when adding 1.50×10^{-3} M of H_2O_2 either in the dark or under light irradiation. The first phenomenon worth mentioning is the so-called "time zero", that is a sample containing all the reagents, filtered immediately after being prepared, and then analyzed. It gave an initial concentration of caffeine from ca. 15 to 25% lower than the theoretic amount (5 mg·L^{-1}), suggesting some H_2O_2 participation to the phenomenon in terms of Fenton reaction. One possible explanation can be correlated with a not perfectly controlled MH0.2 synthetic procedure, in particular it could be ascribed to the unknown hydration extent of $FeCl_3$ used as source of Fe(III) in the synthesis of magnetite; as a result part of the Fe cations can be excluded from the crystalline framework of the magnetite/maghemite phase and affect the reactivity of the prepared material [16]. Indeed this amount of Fe(III) can be more easily available for forming photoactive complexes with the BBS-C moiety, thus justifying an initial very fast caffeine degradation

Moreover, the three HNMPs showed different efficiencies and the process was in all cases faster under irradiation. A test was performed in order to compare the efficiency of the three HMNPs: an aqueous solution of caffeine (5 mg·L^{-1}) was irradiated for 30 min in the presence of HMNPs (200 mg·L^{-1}) and H_2O_2 (1.50×10^{-3} M). The caffeine abatement in the presence of MH0.1, MH0.2, and MH1.0 was 19.5%, 65.0%, and 24.5%, respectively. Based on these results, MH0.2 was used throughout the rest of the work. Figure 4 shows the degradation profiles obtained in the presence of MH0.2 and hydrogen peroxide, either in the dark or under irradiation. As can be observed, the heterogeneous Fenton process yielded slight caffeine abatement in the first 30 min of contact and then remains stable, with an overall caffeine abatement of ca. 20%.

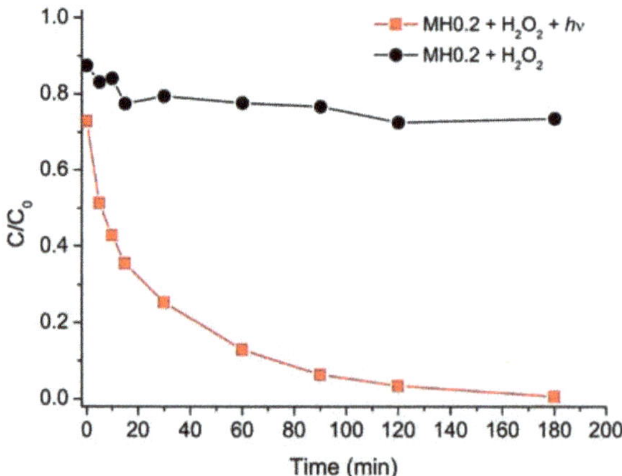

Figure 4. Caffeine relative degradation vs. irradiation time in the presence of MH0.2 (200 mg·L^{-1}) and H_2O_2 (1.50×10^{-3} M). Data reported considering an initial caffeine concentration of 5 mg·L^{-1}. Dark condition (black circles), light irradiation (red squares).

Based on these results no more tests were performed in the dark.

The caffeine degradation profile obtained under irradiation in the presence of MH0.2 and H_2O_2 was then compared with data obtained by irradiating an aqueous caffeine solution in the presence of (i) bare magnetite (M0) and H_2O_2 1.50×10^{-3} M, or (ii) H_2O_2 1.50×10^{-3} M alone. As can be seen in Figure 5, in cases (i) and (ii), almost the same kinetic was obtained, allowing to hypothesize that the degradation process was mainly driven by the photolysis of H_2O_2 yielding hydroxyl radicals (HO·) responsible for the caffeine degradation. On the other side, the beneficial effect of having MH0.2 instead of bare magnetite clearly appears, thus allowing to envisage an active role of the organic moiety.

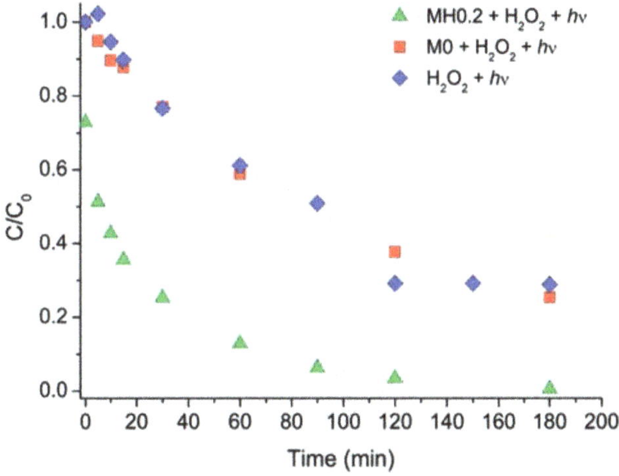

Figure 5. Caffeine relative photodegradation vs. irradiation time in the presence of: H_2O_2 (1.50×10^{-3} M) (blue diamonds); bare magnetite (M0, 200 mg·L^{-1}) and H_2O_2 (1.50×10^{-3} M) (red squares); MH0.2 (200 mg·L^{-1}) and H_2O_2 (1.50×10^{-3} M) (green triangles). Data reported considering an initial caffeine concentration of 5 mg·L^{-1}.

3.5. Effect of pH on the Photo-Fenton Caffeine Degradation Mediated by HMNPs

Since it has been reported that photo-Fenton (and related) processes have their optimum pH at about 2.8, the effect of this experimental variable was checked, aiming to raise this value toward neutrality. We started from pH = 4.5, which was the value measured in the suspension of MH0.2 (200 mg·L^{-1}); such acidic pH could be due to the presence of some hydroxo-Fe(III)-containing phase inside the iron-based MH0.2 core, resulting in Fe(III) hydrolysis. Different experiments were performed rising the initial pH value up to pH = 8.0. As it can be observed in Figure 6, also in the present case the best results could be obtained at lower pH; nevertheless, caffeine degradation was still achievable at pH 6.0 while its degradation rate became too slow at pH 8.0. It is worth to be noted that only at pH 4.5, the concentration at "zero time" was significantly lower than the theoric one. This behavior allows to consider soluble iron, together with the H_2O_2 previously mentioned, responsible for the initial degradation of the caffeine. In fact, only at a very acidic pH can iron species released from the material remain available in solution, whereas at a higher pH they could precipitate in the form of hydroxides.

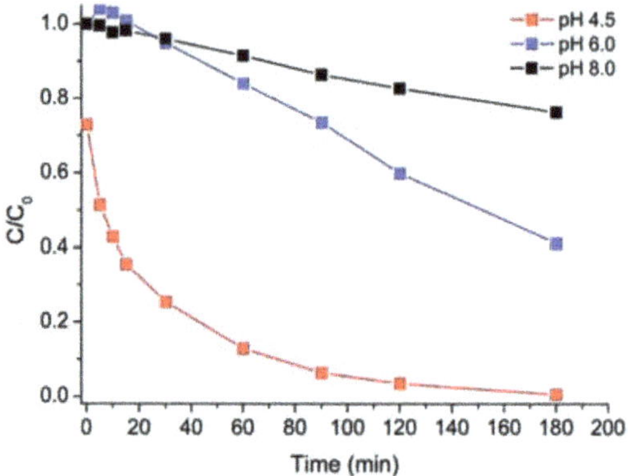

Figure 6. Caffeine relative photodegradation vs. irradiation time at different initial pH values: 4.5 (red squares), 6.0 (blue squares), and 8.0 (black squares) MH0.2 200 mg·L^{-1}, H$_2$O$_2$ 1.50 × 10^{-3} M. Data reported considering an initial caffeine concentration of 5 mg·L^{-1}.

3.6. Effect of Hydroxyl Radical Scavenger and Dissolved Oxygen on the Caffeine Photodegradation Mediated by HMNPs

In order to get insight in the caffeine photodegradation mechanism, experiments have been performed in the presence of 2-propanol that is a HO·scavenger [44]; increasing amounts were added (i.e., 7.5 × 10^{-5} M, 1.0 × 10^{-3} M, and 5.0 × 10^{-3} M), and chosen in order to have a theoric inhibition of caffeine degradation of 50%, 97%, and 100%, respectively. The results reported in Figure 7 show a progressive inhibition effect of 2-propanol on caffeine photodegradation, even if the inhibition did not reach the expected extent in the first two cases. Nevertheless, the main role of HO· in the degradation process was clearly shown. Moreover experiments have been performed by controlling the reaction atmosphere by saturating the system with oxygen or with argon (data not shown); in the absence of oxygen, the degradation slowed down and the calculated first-order kinetic constant decreased from 0.0293 min^{-1} (in air/oxygen) to 0.0164 min^{-1} (in argon-saturated medium), thus confirming the predominant role of oxygenated reactive species as drivers of the caffeine degradation process.

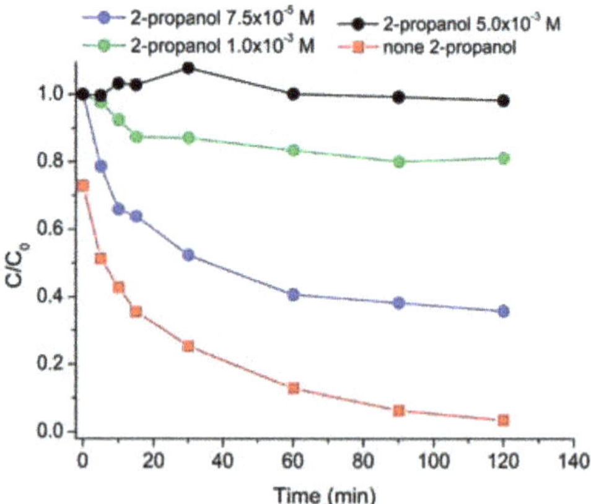

Figure 7. Caffeine relative photodegradation vs. irradiation time in the presence of different [2-propanol]: none 2-propanol (red squares), 7.5 × 10^{-5} M (blue circles), 1.0 × 10^{-3} M (green circles), and 5.0 × 10^{-3} (black circles). MH0.2 200 mg·L^{-1}, H$_2$O$_2$ 1.50 × 10^{-3} M. Data reported considering an initial caffeine concentration of 5 mg L^{-1}.

3.7. Re-Use of HMNPs

In order to consider the possibility to apply HNMPs in real wastewater treatment, an essay of recovery and reuse of MH0.2 was performed. After the first 180 min contact with caffeine under irradiation, MH0.2 were magnetically separated from the aqueous phase; further on, a fresh caffeine solution was added together with hydrogen peroxide and then submitted to irradiation for 180 min. Figure 8 shows the obtained results.

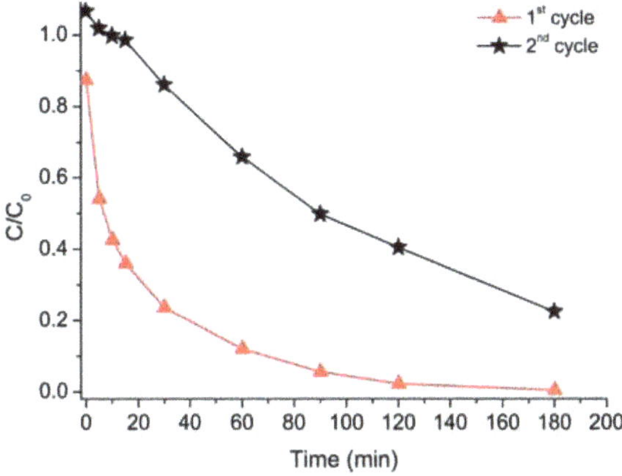

Figure 8. Caffeine relative concentration vs. irradiation time. 1st cycle (red triangles); 2nd cycle (black stars). MH0.2 200 mg·L^{-1}, H$_2$O$_2$ 1.50 × 10^{-3} M. Data reported considering an initial caffeine concentration of 5 mg·L^{-1}.

Several remarkable features could be observed: (i) an overall decrease in the caffeine degradation rate from the first to the second cycle, (ii) an initial 20 min of "induction" step during the second cycle while in the first one a very fast caffeine decay took place in the first 2 min, (iii) the "time zero" effect is visible only in the first cycle.

4. Discussion

The obtained results can be discussed by considering previous data reported in the literature on the use of magnetite as catalyst for photo-Fenton-like pollutant degradation. The negligible caffeine degradation observed under Fenton-like conditions is in agreement with the results reported for phenol degradation in the presence of different magnetite samples [32] and for Bisphenol A degradation in the presence of magnetite and ethylenediamine-*N*,*N*′-disuccinic acid (EDDS) [45], where the irradiation of magnetite with H$_2$O$_2$ was required to trigger reactivity. In the present work, the need of light to induce caffeine transformation and the fact that the relevance of irradiation was not related only to the photolysis of H$_2$O$_2$ allows for the hypothesis that Fe(III) photoreduction to Fe(II) was required to trigger the degradation process. Moreover, the enhanced efficiency observed when comparing MH0.2 towards photo-Fenton-like caffeine abatement with data obtained using bare magnetite was attributed to the presence of the organic BBS-C shell. A first reason could be that the organic shell, behaving as surfactant, enhanced the particle dispersion, yielding higher solid–liquid interface and higher reactivity. Hence, as demonstrated by the effect of added 2-propanol, hydroxyl radicals are mostly responsible for MH0.2 activity, and such species being highly reactive, it is reasonable to hypothesize that the caffeine degradation takes place at the nanoparticles-solution interface rather than in the bulk. A second reason could be that the organic moiety is able to form photoactive complexes with iron, in analogy with the findings (and related interpretation) of Huang and co-workers [45]. The authors stated that EDDS, as a strong chelating agent, maintained iron in soluble form, through Fe-EDDS complex, which is responsible for the superoxide radical anion formation, able to enhance the generation of Fe(II) (the rate-limiting step), and therefore the production of HO·. This implies the consideration also in the present work of the possible role of superoxide anion. In this context, BBS-C that should be able to generate reactive species under light excitation, as shown for other humic-like substances extracted from compost, could contribute to the formation of HO· and/or superoxide radical anion [46,47].

On the other side, a contribution to caffeine photodegradation related to iron and organic matter released in solution (even if at very low concentration), yielding a homogeneous photo-Fenton process cannot be ruled out. In fact, the sample as prepared and used in the photo-Fenton reaction shows the "time zero" phenomenon and a higher activity with respect to the activity of the sample recovered and reused (Figure 8). This behavior could be explained considering that the hydroxo-Fe(III)-containing phase previously mentioned causes the decrease of the pH for hydrolysis, giving soluble iron species active in homogeneous photo-Fenton process for caffeine abatement.

Scheme 2 summarizes the hypothesized mechanism yielding the production of the reactive species (mainly hydroxyl radicals) responsible for caffeine degradation.

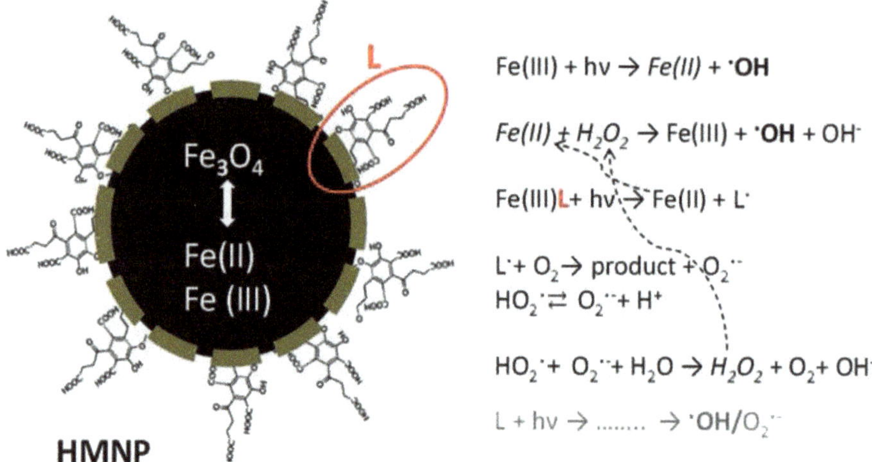

Scheme 2. Schematic representation summarizing the main reactions occurring in the investigated system.

The pH decrease upon the HMNPs' addition, together with a not negligible iron ion released in solution, could contribute to the explanation for the decrement in caffeine abatement efficiency observed in the 2nd cycle, when MH0.2 was reused. Indeed, in the first cycle operated at pH = 4.5, both heterogeneous and homogeneous processes could occur. On the other hand, when MH0.2 were recovered and re-suspended, a slightly higher and less favorable pH (ca. 5.8–6.0) was observed, allowing for the hypothesis that in the second cycle the hydroxo-phase was already eliminated and only the heterogeneous photo-Fenton reaction took place. Even if the efficiency in the second cycle was lower, it is worth mentioning that the photo-Fenton-like reaction was still able to degrade caffeine efficiently, and in much less drastic conditions, since the working pH in the following experiment (i.e., in the absence of Fe(III) hydrolysis) was near the neutrality and this situation, together with the possibility of recovering the material thanks to its magnetism, makes the process potentially with no impact on the environment.

5. Conclusions

The use of HMNPs clearly evidenced the possibility of attaining pollutant degradation working at pH values closer to neutrality, overcoming the limit of pH = 2.8 characteristic of the homogeneous photo-Fenton process. Moreover, the added value of the organic coverage of the nanoparticles was demonstrated, thus allowing for the possibility to exploit organic waste as source of auxiliaries in environmental processes. Nevertheless, further investigation is needed to better understand the reaction, and in particular, to assess the capability of BBS-Cs to complex iron, in either Fe(II) or Fe(III) form.

Lastly, the good preliminary results obtained when reusing the nanoparticles encourage further studies in the direction of scaling-up the process to pilot plant for real application purposes.

Author Contributions: A.B.P. and M.B. conceived and designed the experiments; D.P., R.N. and D.F. performed the experiments; A.B.P., G.M. (Giuliana Magnacca), G.M. (Gilles Mailhot), D.P., C.R. and M.B. analyzed the data; D.F. and G.M. (Giuliana Magnacca) contributed reagents/materials/analysis tools; A.B.P., R.N., M.B. and D.P. wrote the paper.

Funding: This research was funded by the Marie Sklodowska-Curie Research and Innovation Staff Exchange project funded by the European Commission grant number 645551 (Mat4teaT) and 765860 (AQUAlity). This research was funded by Compagnia di San Paolo and University of Torino (Microbusters). This research was funded by Polytechnic of Torino grant number 54_RSG17NIR01 (Starting Grant RTD).

Acknowledgments: This work was realized with the financial support for academic interchange by the Marie Sklodowska-Curie Research and Innovation Staff Exchange project funded by the European Commission H2020-MSCA-RISE-2014 within the framework of the research project Mat4treaT (Project number: 645551). This paper is part of a project that has received funding from the European Union's Horizon 2020 research and innovation programme under the Marie Sklodowska-Curie grant agreement No. 765860 (AQUAlity). Compagnia di San Paolo and University of Torino are gratefully acknowledged for funding Project Torino_call2014_L2_126 through "Bando per il finanziamento di progetti di ricerca di Ateneo—anno 2014" (Project acronym: Microbusters). Polytechnic of Torino is gratefully acknowledged for funding project Starting Grant RTD (project number: 54_RSG17NIR01). M.B. and G. Mai acknowledge financial support from the CAP 20-25 I-site project.

Conflicts of Interest: The authors declare no conflict of interest. The founding sponsors had no role in the design of the study; in the collection, analyses, or interpretation of data; in the writing of the manuscript, and in the decision to publish the results.

References

1. United Nations General Assembly (UNGA). *The Human Right to Water and Sanitation*; Resolution 64/292; United Nations: New York, NY, USA, 2010.
2. United Nations General Assembly (UNGA). *Transforming Our World: The 2030 Agenda for Sustainable Development*; United Nations: New York, NY, USA, 2015.
3. Clean Water and Sanitation. Available online: http://www.un.org/sustainabledevelopment/water-and-sanitation/ (accessed on 1 March 2018).
4. Applying the Circular Economy Lens to Water. Available online: http://circulatenews.org/2017/01/applying-the-circular-economy-lens-to-water/ (accessed on 1 March 2018).
5. Yang, Y.; Ok, Y.S.; Kim, K.-H.; Kwon, E.E.; Tsang, Y.F. Occurrences and removal of pharmaceuticals and personal care products (PPCPs) in drinking water and water/sewage treatment plants: A review. *Sci. Total Environ.* **2017**, *596*, 303–320. [CrossRef] [PubMed]
6. Tiedeken, E.J.; Tahar, A.; McHugh, B.; Rowan, N.J. Monitoring, sources, receptors, and control measures for three European Union watch list substances of emerging concern in receiving waters—A 20year systematic review. *Sci. Total Environ.* **2017**, *574*, 1140–11673. [CrossRef] [PubMed]
7. Barbosa, M.O.; Moreira, N.F.F.; Ribeiro, A.R.; Pereira, M.F.R.; Silva, A.M.T. Occurrence and removal of organic micropollutants: An overview of the watch list of EU Decision 2015/495. *Water Res.* **2016**, *94*, 257–279. [CrossRef] [PubMed]
8. Salimi, M.; Esrafili, A.; Gholami, M.; Jonidi Jafari, A.; Rezaei Kalantary, R.; Farzadkia, M.; Kermani, M.; Sobhi, H.R. Contaminants of emerging concern: A review of new approach in AOP technologies. *Environ. Monit. Assess.* **2017**, *189*, 414. [CrossRef] [PubMed]

9. Bernabeu, A.; Vercher, R.F.; Santos-Juanes, L.; Simón, P.J.; Lardín, C.; Martínez, M.A.; Vicente, J.A.; Gonzalez, R.; Llosa, C.; Arques, A.; et al. Solar photocatalysis as a tertiary treatment to remove emerging pollutants from wastewater treatment plant effluents. *Catal. Today* **2011**, *161*, 235–240. [CrossRef]

10. Borowska, E.; Bourgin, M.; Hollender, J.; Kienle, C.; McArdell, C.S.; Von Gunten, U. Oxidation of cetirizine, fexofenadine and hydrochlorothiazide during ozonation: Kinetics and formation of transformation products. *Water Res.* **2016**, *94*, 350–362. [CrossRef] [PubMed]

11. Bouafia-Chergui, S.; Zemmouri, H.; Chabani, M.; Bensmaili, A. TiO_2-photocatalyzed degradation of tetracycline: Kinetic study, adsorption isotherms, mineralization and toxicity reduction. *Desalin. Water Treat.* **2016**, *57*, 16670–16677. [CrossRef]

12. Casas, M.E.; Bester, K. Can those organic micro-pollutants that are recalcitrant in activated sludge treatment be removed from wastewater by biofilm reactors (slow sand filters)? *Sci. Total Environ.* **2015**, *506–507*, 315–322. [CrossRef] [PubMed]

13. Nisticò, R.; Franzoso, F.; Cesano, F.; Scarano, D.; Magnacca, G.; Parolo, M.E.; Carlos, L. Chitosan-derived iron oxide systems for magnetically guided and efficient water purification processes from polycyclic aromatic hydrocarbons. *ACS Sustain. Chem. Eng.* **2017**, *5*, 793–801. [CrossRef]

14. De la Cruz, N.; Esquius, L.; Grandjean, D.; Magnet, A.; Tungler, A.; de Alencastro, L.F.; Pulgarin, C. Degradation of emergent contaminants by UV, UV/H_2O_2 and neutral photo-Fenton at pilot scale in a domestic wastewater treatment plant. *Water. Res.* **2013**, *47*, 5836–5845. [CrossRef] [PubMed]

15. Demarchis, L.; Minella, M.; Nisticò, R.; Maurino, V.; Minero, C.; Vione, D. Photo-Fenton reaction in the presence of morphologically controlled hematite as iron source. *J. Photochem. Photobiol. A* **2015**, *307–308*, 99–107. [CrossRef]

16. Nisticò, R. Magnetic materials and water treatments for a sustainable future. *Res. Chem. Intermed.* **2017**, *43*, 6911–6949. [CrossRef]

17. Klamerth, N.; Malato, S.; Maldonado, M.I.; Aguera, A.; Fernández-Alba, A.R. Application of photo-Fenton as a tertiary treatment of emerging contaminants in municipal wastewater. *Environ. Sci. Technol.* **2010**, *44*, 1792–1798. [CrossRef] [PubMed]

18. Klamerth, N.; Rizzo, L.; Malato, S.; Maldonado, M.I.; Aguera, A.; Fernández-Alba, A.R. Degradation of fifteen emerging contaminants at μgL^{-1} initial concentrations by mild solar photo-Fenton in MWTP effluents. *Water Res.* **2010**, *44*, 545–554. [CrossRef] [PubMed]

19. Bauer, R.; Fallmann, H. The photo-Fenton oxidation—A cheap and efficient wastewater treatment method. *Res. Chem. Intermed.* **1997**, *23*, 341–354. [CrossRef]

20. Wu, Y.L.; Passananti, M.; Brigante, M.; Dong, W.B.; Mailhot, G. Fe(III)-EDDS complex in Fenton and photo-Fenton processes: From the radical formation to the degradation of a target compound. *Environ. Sci. Pollut. Res.* **2014**, *21*, 12154–12162. [CrossRef] [PubMed]

21. Montoneri, E.; Mainero, D.; Boffa, V.; Perrone, D.G.; Montoneri, C. Biochemenergy: A project to turn an urban wastes treatment plant into biorefinery for the production of energy, chemicals and consumer's products with friendly environmental impact. *Int. J. Glob. Environ. Issues* **2011**, *11*, 170–196. [CrossRef]

22. Palma, D.; Bianco Prevot, A.; Celi, L.; Martin, M.; Fabbri, D.; Magnacca, G.; Chierotti, M.R.; Nisticò, R. Isolation, characterization, and environmental application of bio-based materials as auxiliaries in photocatalytic processes. *Catalysts* **2018**, *8*, 197. [CrossRef]

23. Nisticò, R.; Barrasso, M.; Carrillo Le Roux, G.A.; Seckler, M.M.; Sousa, W.; Malandrino, M.; Magnacca, G. Biopolymers from composted biowaste as stabilizers for the synthesis of spherical and homogeneously sized silver nanoparticles for textile applications on natural fibers. *ChemPhysChem* **2015**, *16*, 3902–3909. [CrossRef] [PubMed]

24. Avetta, P.; Bella, F.; Bianco Prevot, A.; Laurenti, E.; Montoneri, E.; Arques, A.; Carlos, L. Waste cleaning waste: Photodegradation of monochlorophenols in the presence of waste derived organic catalysts. *ACS Sustain. Chem. Eng.* **2013**, *1*, 1545–1550. [CrossRef]

25. Montoneri, E.; Bianco Prevot, A.; Avetta, P.; Arques, A.; Carlos, L.; Magnacca, G.; Laurenti, E.; Tabasso, S. Food wastes conversion to products for use in chemical and environmental technology, material science and agriculture. In *The Economic Utilisation of Food Co-Products*; Kazmi, A., Shuttleworth, P., Eds.; Royal Society of Chemistry Publishing: Cambridge, UK, 2013; pp. 64–109. ISBN 978-1-84973-615-2.

26. Nisticò, R.; Cesano, F.; Franzoso, F.; Magnacca, G.; Scarano, D.; Funes, I.G.; Carlos, L.; Parolo, M.E. From biowaste to magnet-responsive materials for water remediation from polycyclic aromatic hydrocarbons. *Chemosphere* **2018**, *202*, 686–693. [CrossRef] [PubMed]

27. What is Circular Economy? Available online: https://www.ellenmacarthurfoundation.org/circular-economy (accessed on 10 March 2018).

28. Gomis, J.; Bianco Prevot, A.; Montoneri, E.; Gonzalez, M.C.; Amat, A.M.; Martire, D.O.; Arques, A.; Carlos, L. Waste sourced bio-based substances for solar-driven wastewater remediat ion: Photodegradation of emerging pollutants. *Chem. Eng. J.* **2014**, *235*, 236–243. [CrossRef]

29. Gomis, J.; Carlos, L.; Bianco Prevot, A.; Teixeira, A.C.S.C.; Mora, M.; Amat, A.M.; Vicente, R.; Arques, A. Bio-based substances from urban waste as auxiliaries for solar photo-Fenton treatment under mild conditions: Optimization of operational variables. *Catal. Today* **2015**, *240*, 39–45. [CrossRef]

30. Pignatello, J.J.; Oliveros, E.; MacKay, A. Advanced oxidation processes for organic contaminant destruction based on the Fenton reaction and related chemistry. *Crit. Rev. Environ. Sci. Technol.* **2006**, *36*, 1–84. [CrossRef]

31. Nisticò, R.; Celi, L.R.; Bianco Prevot, A.; Carlos, L.; Magnacca, G.; Zanzo, E.; Martin, M. Sustainable magnet-responsive nanomaterials for the removal of arsenic from contaminated water. *J. Hazard. Mater.* **2018**, *342*, 260–269. [CrossRef] [PubMed]

32. Minella, M.; Marchetti, G.; De Laurentiis, E.; Malandrino, M.; Maurino, V.; Minero, C.; Vione, D.; Hanna, K. Photo-Fenton oxidation of phenol with magnetite as iron source. *Appl. Catal. B Environ.* **2014**, *154–155*, 102–109. [CrossRef]

33. Li, Y.; Yuan, D.; Dong, M.; Chai, Z.; Fu, G. Facile and green synthesis of core-shell structured magnetic chitosan submicrospheres and their surface functionalization. *Langmuir* **2013**, *29*, 11770–11778. [CrossRef] [PubMed]

34. Magnacca, G.; Allera, A.; Montoneri, E.; Celi, L.; Benito, D.E.; Gagliardi, L.G.; González, M.C.; Mártire, D.O.; Carlos, L. Novel magnetite nanoparticles coated with waste-sourced biobased substances as sustainable and renewable adsorbing materials. *ACS Sustain. Chem. Eng.* **2014**, *2*, 1518–1524. [CrossRef]

35. Bianco Prevot, A.; Baino, F.; Fabbri, D.; Franzoso, F.; Magnacca, G.; Nisticò, R.; Arques, A. Urban biowaste-derived sensitizing materials for caffeine photodegradation. *Environ. Sci. Pollut. Res.* **2017**, *24*, 12599–12607. [CrossRef] [PubMed]

36. Franzoso, F.; Nisticò, R.; Cesano, F.; Corazzari, I.; Turci, F.; Scarano, D.; Bianco Prevot, A.; Magnacca, G.; Carlos, L.; Martire, D.O. Biowaste-derived substances as a tool for obtaining magnet-sensitive materials for environmental applications in wastewater treatments. *Chem. Eng. J.* **2017**, *310*, 307–316. [CrossRef]

37. Dafouz, R.; Cáceres, N.; Rodríguez-Gil, J.L.; Mastroianni, N.; López de Alda, M.; Barceló, D.; Gil de Miguel, A.; Valcárcel, Y. Does the presence of caffeine in the marine environment represent an environmental risk? A regional and global study. *Sci. Total Environ.* **2018**, *615*, 632–642. [CrossRef] [PubMed]

38. Florawiva Compost di Qualità. Available online: http://ambiente.aceapinerolese.it/Florawiva_prese.html (accessed on 24 November 2017).

39. Cesano, F.; Fenoglio, G.; Carlos, L.; Nisticò, R. One-step synthesis of magnetic chitosan polymer composite films. *Appl. Surf. Sci.* **2015**, *345*, 175–181. [CrossRef]

40. Sodano, M.; Lerda, C.; Nisticò, R.; Martin, M.; Magnacca, G.; Celi, L.; Said-Pullicino, D. Dissolved organic carbon retention by coprecipitation during the oxidation of ferrous iron. *Geoderma* **2017**, *307*, 19–29. [CrossRef]

41. Peuravuori, J.; Pihlaja, K. Molecular size distribution and spectroscopic properties of aquatic humic substances. *Anal. Chim. Acta* **1997**, *337*, 133–149. [CrossRef]

42. Franzoso, F.; Vaca-Garcia, C.; Rouilly, A.; Evon, P.; Montoneri, E.; Persico, P.; Mendichi, R.; Nisticò, R.; Francavilla, M. Extruded versus solvent cast blends of poly(vinyl alcohol-*co*-ethylene) and biopolymers isolated from municipal biowaste. *J. Appl. Polym. Sci.* **2016**, *133*, 43009. [CrossRef]

43. Avetta, P.; Berto, S.; Bianco Prevot, A.; Minella, M.; Montoneri, E.; Persico, D.; Vione, D.; Gonzalez, M.C.; Mártire, D.O.; Carlos, L.; et al. Photoinduced transformation of waste-derived soluble bio-based substances. *Chem. Eng. J.* **2015**, *274*, 247–255. [CrossRef]

44. Vione, D.; Maurino, V.; Minero, C.; Pelizzetti, E. Phenol photonitration upon UV irradiation of nitrite in aqueous solution I: Effects of oxygen and 2-propanol. *Chemosphere* **2001**, *45*, 893–902. [CrossRef]

45. Huang, W.Y.; Luo, M.Q.; Wei, C.S.; Wang, Y.H.; Hanna, K.; Mailhot, G. Enhanced heterogeneous photo-Fenton process modified by magnetite and EDDS: BPA degradation. *Environ. Sci. Pollut. Res.* **2017**, *24*, 10421–10429. [CrossRef] [PubMed]

46. Amine-Khodja, A.; Trubetskaya, O.; Trubetskoj, O.; Cavani, C.; Ciavatta, C.; Guyot, G.; Richard, C. Humic-like substances extracted from composts can promote the photodegradation of Irgarol 1051 in solar light. *Chemosphere* **2005**, *62*, 1021–1027. [CrossRef] [PubMed]

47. Calza, P.; Di Sarro, J.; Magnacca, G.; Bianco Prevot, A.; Laurenti, E. Low-cost magnetic materials containing waste derivatives as catalyst for removal of organic pollutants: Insights into the reaction mechanism and odd aspects. In Proceedings of the 6th International Conference on Sustainable Solid Waste Management, Naxos Island, Greece, 13–16 June 2018.

 © 2018 by the authors. Licensee MDPI, Basel, Switzerland. This article is an open access article distributed under the terms and conditions of the Creative Commons Attribution (CC BY) license (http://creativecommons.org/licenses/by/4.0/).

MDPI
St. Alban-Anlage 66
4052 Basel
Switzerland
Tel. +41 61 683 77 34
Fax +41 61 302 89 18
www.mdpi.com

Materials Editorial Office
E-mail: materials@mdpi.com
www.mdpi.com/journal/materials

www.ingramcontent.com/pod-product-compliance
Lightning Source LLC
LaVergne TN
LVHW071521180726
843512LV00014B/1134

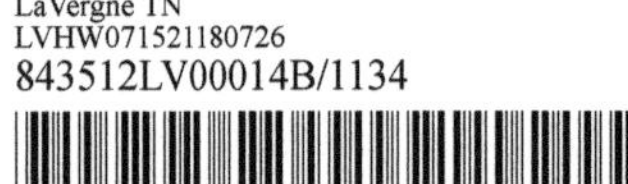